The

Human

Hologram

Living Your Life in Harmony with the Unified Field

Dr. Robin Kelly

www.TheHumanHologram.com

Energy Psychology Press
Santa Rosa, CA 95439
www.EnergyPsychologyPress.com

Library of Congress Cataloging-in-Publication Data

Kelly, Robin, 1951—
   The human hologram: living your life in harmony with the unified field / Robin Kelly. —
1st ed.
       p. cm.
Includes bibliographical references and index.
ISBN: 978-1-60415-062-9 (pbk.)
1. Holistic medicine. 2. Holography. 3. Three-dimensional imaging.
4. Mind and body. I. Title.

R733.K454 2011
610—dc23

2011013519

© 2011 Robin Kelly

Typeset in Adobe Garamond Pro & Lucida Sans by Karin Kinsey
Printed in USA by Bang Printing
1st Edition

10 9 8 7 6 5 4 3 2 1

For

Pamela Mary Kelly

April 22, 1920–January 1, 2010

"It's not as simple as that, Robin."

Notices

Figures 1, 12, 14, and 17 are adapted from illustrations in *The Human Antenna* by Robin Kelly.

Figure 28 is adapted from an illustration in *Healing Ways* by Robin Kelly.

Part of the text in chapter 10 appears in *The Human Antenna* (Appendix 1.)

All reproduced with permission.

Contents

Acknowledgments

The Human Hologram would never have been conceived or allowed to evolve without the enthusiastic support of the great team at Elite Books and Energy Psychology Press. I am grateful for the vision and guidance of Dawson Church, the expertise and patience of Courtney Arnold and Deb Tribbey, and the wisdom and artistry of Stephanie Marohn and Karin Kinsey. And here in New Zealand, I must thank Ruth Hamilton for her time and expertise shared so generously, and Bruce Lipton for his valuable advice, his pioneering work, and his inspiring presence each summer.

To those whose life stories are recounted here, thank you so much. I hope I have captured your essence and truthfully conveyed the valuable lessons you have taught me over the years.

Thanks as ever to my trusted friends Gerald and Miriam Gibb, and Tim and Julie Ewer who continue to stretch and soften the hide that binds our noble profession so tightly. Your encouragement, advice, and feedback sustained me throughout this project.

And to Trish and my family—you are simply the best.

Introduction

A little over seven hundred years ago, in the northern Italian town of Padua, a young artist was making history. It was not the subject matter that was to set Giotto di Bondone apart from all who had painted before him; his series of frescoes that still adorn the walls of Scrovegni Chapel depicted the lives and times of Jesus and the Virgin Mary. What made Giotto so special was the revolutionary three-dimensional quality of his works—vivid images that invited viewers into the heart of each scene. To the people of the day, he made the figures of the disciples, and even of Jesus and Mary, seem strangely familiar, just like them and their friends. The divine was no longer so distant, now more present than ever in their lives.

Giotto even added bold contemporary touches: the Star of Bethlehem in his nativity scene appears clearly as a fiery amber comet. It is known that Halley's comet would have been visible from Earth in October 1301, shortly before these frescoes were commissioned. In 1986, the European spacecraft that successfully ventured within six hundred kilometers of Halley's comet in 1986 was to take his name.

As the first Western artist to use perspective in his work, he is now regarded as the true father of Renaissance art. It would be nearly two centuries later that others, Michelangelo, Leonardo, and Raphael amongst them, would take this exciting new vision of humanity and the cosmos to the world. Through their artistic vision and practical skills, they were to forge lasting bonds between science and spirituality that are being rediscovered today through the popular writings of Dan Brown, Michael Baigent, and many others.

It is perhaps strange that the three-dimensional worldview that we today take for granted could have been so difficult to capture on canvas or stone before the early fourteenth century. Nowadays, children as young as seven at primary school rapidly master the art of drawing cubes, houses, and castles using the laws of perspective. Yet it took the creativity of a small group of individuals in the Middle Ages to free up the consciousness of the Western world.

And now, in the early years of the twenty-first century, we are witnessing an expansion of global consciousness to match, even surpass, that of fourteenth- and fifteenth-century Europe. We are privy to the science, the technology, and the underlying human ingenuity that guide us to appreciate that there may be far more to our lives and our surroundings than the four dimensions that our senses and brains convey.

For more and more movies, I am now donning special polarized or colored glasses that allow selective images to reach each of my eyes. By so manipulating my sense of vision, the blurred and insipid jumbled mess on the screen is instantly transformed into a three-dimensional world, a true virtual reality that has the power to involve and move me like never before. I witness on YouTube a virtual holographic image of Al Gore addressing the 2007 Live Earth conference in Tokyo. I also see the three-dimensional image of a lecturer being transmitted, instantly in "real time," into a distant lecture hall full of students. What is more, the lecture hall, including the students, is simultaneously transmitted back to him, allowing the vibrant free-flow of interaction so missing from previously sterile teleconferencing. There is even technology now that allows us to feel with our fingers the ghost-like image projected in our midst.

It is staggering to imagine how far this virtual academic world will go. At boarding school, I was once the ungrateful recipient of a hard wooden blackboard eraser aimed with great accuracy at my head by a history teacher somehow finding fault with my efforts to concentrate on his words of wisdom. It will likely, I suspect, be modern codes of teachers' conduct, rather than advancing technology, that will save future generations of daydreaming pupils from such a painful fate.

But new scientific insights are taking us far beyond the worlds of communication, entertainment, and even education. In recent years, the leading popular scientific magazines from each side of the Atlantic have had as their cover stories the startling hypothesis that we are ourselves are, at our very core, holographic beings, living within a holographic universe. That the three- and four-dimensional world of ourselves, our homes, our Toyotas, and our mountains is merely a projection—a virtual reality created and perceived by our five senses. Our universe, in reality, some scientists claim, may be more like a giant flat movie screen, a cosmic Imax, consisting of trillion upon trillion of pixels, converted by our collective living gaze into our past, present, and future surroundings—and yes, even into our own selves.

It is tempting, and I dare say many will say more sensible, to stand aside from this theory, dismissing it as being altogether irrelevant and "way too crazy." Without a doubt, too much information. Let's not be sidetracked by this weirdness. Rather, let's just get on with our lives, facing up to the huge challenges we all face today—global financial chaos, climate change, terrorism, and nuclear weapon proliferation.

I understand this reaction. These were my thoughts entirely … initially. But then questions emerged within me that began to nag. What do we know of the science of holograms that might shed light on the human condition? Does the fact that every part of the special photographic plate used in the creation of a holographic image contains the information of the whole have any relevance to our own bodies? What use would it to be to us and our planet if, indeed, it was true that our observations and actions were intricately, and intimately, involved in the workings of our universe?

The second section of this book explores the science of the hologram. But in the manner of Giotto di Bondone and the magical folk at Pixar, I wish to draw you inside this brave new scientific world as a fully involved enthusiastic observer. An observer that helps create the observed—the singer, at the same time, the song. I am keen for you to perceive that your unique understanding of what follows in this book will make it easier for all of us to grasp concepts that seem just beyond our reach. Your wisdom will expand the wisdom of all.

I will stay clear from complex scientific formulae, for one simple reason—I struggle to understand them. But I am immensely grateful to those who do, for it is their hard work and intellect that is as wondrous to me as any of the insights that might emerge for us during the writing and reading of this book. I am fully aware from the feedback from my previous books that, to many, the word "science" equates with "Oh heck, this is going to be hard." If this is your instant reaction, I can suggest two exercises that may help. First, try tapping on your chest while saying these words: "Even though the word 'science' reminds me of a subject at school I hated, and now makes me feel inadequate, I really love and respect myself for the many and varied wonderful skills I have brought into this world."

If, after doing this on three consecutive occasions, the word "science" still induces a state of lingering inadequacy and mild terror, it's time to go directly to Plan B. This one is even easier: simply substitute the word "science" whenever it appears in this book or comes into your head with an alternative word, which to you summons delicious thoughts and feelings. "Chocolate," "chardonnay," or even "cheesecake" might work.

For this is a science that cannot be understood using our intellectual minds alone. It has first to be felt, recognized, and welcomed with open arms as if we were greeting a long-lost friend. An old friend who then begins to astound us with stories of herself that we have never heard before. And who somehow knows something wonderful about us that has been kept hidden all our lives.

The third section of this book explores how an understanding of ourselves as human holograms, and as players in the great unified field of consciousness, impacts on our day-to-day lives. As a family doctor of thirty years, every day I am brought down to earth by the real, immediate needs of people in some degree of minor or major crisis. Collectively, they stretch, baffle, entertain, and enlighten me in their attempts to get better. They come randomly, but purposefully, to my door—folks of all shapes, sizes, and ages. Each with their own set of passions and beliefs. It is this part of my life, and the people that occupy it, that provides the rich material for this middle section, the heart of my book.

Within a year of starting in practice, I was using acupuncture to help facilitate the healing process of patients seeking my help. Although at the time I would have explained it somewhat differently, I now realize that I was being introduced to the holographic paradigm. Studying the Taoist principles of Chinese medicine alongside Western medicine, I slowly became aware of the reality that the human body reflects the universal traits and truths of nature. That the macrocosm is represented within us, the microcosm—"as above, so below."

On a daily basis, I was inserting needles in patients' ears to relieve pain, nausea, and sleeplessness. I was experiencing, at the practical frontline of primary medical care, the fractal reality of a microsystem in which all the information of the whole body can be accessed from one small part, the outer ear.

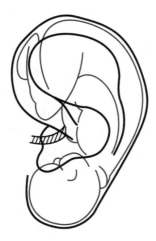

Figure 1. Information of the body in the outer ear. After Dr. P. M. F. Nogier.[1]

In addition, I was learning from Chinese physicians how to gain valuable information about the body by reading patterns on the surface of a human tongue, and by palpating the pulse on the wrist with the tips of my fore, middle, and ring fingers. Twenty-eight years later, we are witnessing a coming of age of holographic medicine, with ear acupuncture being used by American medics and paramedics in Afghanistan and Iraq in the battlefield as a first intervention in controlling the pain of injuries incurred by troops in action.[2] The rapidity of its painkilling effects, as well as its proven safety record, has overridden fears that the science that lies behind its action remains elusive to the majority of the medical profession.

In 1971, Dennis Gabor won the Nobel Prize for his groundbreaking research on holograms in 1947. Like that of Copernicus and Galileo in the sixteenth and seventeenth centuries, his work represents a paradigm-shifting advance in our understanding of the universe, and our role within it. By creating three-dimensional structures from overlapping beams of laser light, his landmark research built on the work of many other twentieth-century physicists who had explored the underlying energetic basis of life. This closely correlated with traditional, empirically based wisdoms and cultures, from the Chinese to Native Americans, with their understanding that the realms of energy, consciousness, and spirit are more fundamental than the material world detected by our five senses. The modern age of computers, cyberspace, and quantum science is guiding us towards an understanding that even more fundamental than energy is the concept of fields of information.

Computer science has allowed us to explore the world of fractals—the patterns of nature that reveal to us that despite the vast diversity of form that surrounds us, and intrigues us so, all matter is fundamentally connected, unified at its source. Health and happiness, I have learned to consider, peaks within us when we are in true harmony with this unified field. Wherever there is disharmony that needs correcting, it is conveyed to us through the feelings, or symptoms, of our body. As children, before our rational brains become fully developed, these feelings need to be acknowledged by our parents and guardians. It is then the parents' responsibility to work out ways to create a harmonious field in which their child can flourish. If this is neglected, it may reflect in episodes of ill health throughout that child's life. Genes associated with many diseases that will lay dormant in a nurtured, and self-nurtured environment, may be turned on or "expressed" when exposed to disharmony in childhood, to be triggered further when issues such as abandonment or abuse recur in later years.

The model of healing I present in the third section is built around the hypothesis that we are connected holographically to a great unifying field of consciousness. The healer, I suggest, adopts a role of parent, listening and feeling first, then engaging his or her rational brain. The compassionate act of listening is followed by the equally compassionate act of the conscious initiation of healing. This process then acts as a template for all future aspects of the life of the healee, who learns to live and love life in the now, in harmony.

This is the framework within which I prefer to conduct my consultations: as a fractal "blueprint" matching a mode of living that is both sustainable and affirming. As in parenting, this process educates the healee into becoming both responsible and independent—the true owner of her or his health.

Appendix 1 contains exercises and hints that will ensure your involvement in the process, if so needed. These can be used on their own or added in to enhance all manner of health consultations with professionals from orthopedic specialists to Reiki practitioners. My own experiences over thirty years strongly suggest to me that this approach to health carries with it not only deep personal advantages for the patient or client, but also short- and long-term economic benefits. It leads to less prescribing of pharmaceuticals and, I am sure, if instigated as a health plan early in life, it will lead to less chronic illness in later life. It is, I believe, the epitome of good sensible medicine.

The fourth section of the book addresses how an awareness of ourselves as human holograms will impact our lives and communities in the broadest sense. The emerging paradigm will enlighten humans about their extraordinary ability to focus their consciousness for the greater good—as individuals and, perhaps most important, collectively. Rational intellect will be seen as being as valuable as ever for our survival, but now only if it is balanced by truly compassionate intent. The holographic human body is recognized as a unit fully integrated both within itself and within the cosmos. The hierarchy of head ruling over heart—intellect over feeling—will be replaced by an understanding of how these vital organs relate and, ideally, cooperate with each other in perfect balance. The old harsh hierarchy will be replaced by a new harmonious holarchy.

The twentieth century—the most violent hundred years in human history—has instilled in us a wariness of cultism. The world witnessed a rational nation hypnotized by one dominant human into performing collective acts of destruction and genocide on a scale never seen before. On a smaller scale, but as devastating to those involved, we have watched cult leaders lead their "disciples" into suicide pacts in Waco and Jonestown. And within this century, brainwashing—a rehypnotizing of the discontented—is as present as ever, with fundamentalist terror groups threatening world peace, and drug and porn barons luring young adults into dangerous and destructive lifestyles.

Compassionate intent—heart consciousness—is, to my mind, the most powerful antidote to these threats, and I am sure holographic science and philosophy with their focus on harmony and balance, can play a major role in guiding us towards a peaceful future.

Similarly, insights into our holographic connections to nature will prove essential as the world unites to reverse climate changes that threaten our future here on Earth. Perhaps it is no coincidence that human consciousness is now at a breakthrough point—a new renaissance—just as we face these threats. We realize that we have to cooperate, join hands, and forgive past indiscretions if we are to have a real chance of thriving on this planet. We realize we are guardians of this Earth and all its life forms. So I would like to think that, through synchronicity and serendipity, the paradigm that embraces our holographic reality has arrived "in the nick of time." For this is a paradigm that recognizes that time itself is relative, a perception to help us achieve and act in the miraculous third and fourth dimensions. Maybe the reason some interpret the Mayan calendar as ending around 2012 is that our perception of time as fixed and linear is changing. So, rather than this date signifying an apocalyptic end of time, it is really and far more optimistically, the end of time-dependence.

There is also a current and controversial theory that suggests there will be an unprecedented leap of human consciousness as our planet's magnetic field aligns with the center of our galaxy. Is it possible, however, that, rather than this being solely a physical process in the cosmos way beyond our control and involvement, it is our very awareness of our personal holographic connections with our earth, our sun, and our galaxy that lies at the heart of this alignment? And that by achieving balance within ourselves, we instantly contribute to balance within our universe? We—you and I—are ultimately responsible.

So thank you for joining me as a fellow explorer in the recently discovered and uncharted territories of this exciting new paradigm. In the following pages, I'll build upon my previous books *Healing Ways* and *The Human Antenna* in the organic way my own life's experiences have grown, layer upon layer. I'll continue to explain my personal journey of discovery for one reason, and one reason only: so that you can relish your own, and fully live out your own perfect, colorful, and unique role. For the underlying message that runs through each of these pages is that it is you, yourself, who owns this responsibility to heal, and to contribute proactively, through benevolent acts to yourself and others, to the unified field of consciousness from which all we know emanates.

To quote Aesop, a humble Greek slave from the sixth century BC: "No act of kindness, no matter how small, is ever wasted."

Section One:
Living in the Holographic Universe

*When pure sincerity forms within, it is outwardly
realized in other people's hearts.*
—Lao Tzu, sixth century BC, Chinese philosopher

It is a beautiful late summer's afternoon. My family is shopping in the local mall. I drive for ten minutes to my favorite beach to swim, relax, and wallow in the calm warm water. I walk to the ocean's edge, submerge, turn, and float on my back. Closing my eyes, I find myself completely at ease, conscious only of being cradled and clothed by a sea that, for today at least, only nurtures. I reflect on John F. Kennedy's speech to the America's Cup crews in 1962, thinking that maybe, as he said, this is "because we all came from the sea... that all of us have in our veins the exact same percentage of salt in our blood that exists in the ocean...and when we go back to the sea...we are going back from whence we came."

But really, I am not thinking that much, merely focusing on being. Today there are no waves to rock my world; I am in a state of suspended animation, away from daily pressures, oblivious to the passage of time. Refreshed and at peace, I make my way back to the sandy beach, dry myself, and change into my T-shirt and shorts. I turn round to find a young girl, possibly only three years old, staring up at me. She isn't crying or visibly distressed.

"Hello there," I say, looking around for her mother or father. But the only people on the beach are two teenage boys about fifty meters away. "Is she with you?" I shout. "Definitely not!" one yells back, as if the very suggestion were entirely ridiculous, not to mention insulting.

I look farther afield. Between the beach and the road there is a small play area—a slide, some swings, a roundabout, and a couple of picnic tables. But it is completely deserted. I ask the little girl for her name and she hides coyly behind her hands.

"Follow me," I say. She doesn't move. When I offer her my hand, she shrugs and moves away.

I have a dilemma. Do I pick her up and go searching for her parents, relieving them of their obvious anguish as rapidly as possible, or just stay with her here, keeping her safe while waiting for someone to find her? Suddenly, I feel very vulnerable—will it take some explaining if I, a middle-aged stranger, appear to the parents out of blue clutching their three-year-old daughter? Maybe better to stay where I am.

My afternoon was proving to be far less peaceful than I had planned. I wait for five minutes, scanning the beach and surroundings for family members, listening intently for someone calling out a child's name. Then from somewhere far behind me, at the opposite end of the beach, comes a male voice, "Hayley…Hayley," and then a female voice calling the same. Soon after, a couple, visibly distressed, emerge from a small thicket of young pines. They spot their daughter and me, I wave to them, pick up their daughter, and run towards them.

In this instant, any complexities or worrying implications of my tricky situation are dismissed, as I respond instinctively to an urgent need to reunite a lost little girl with her frantic parents. I let Hayley down onto the sand as we approach each other, and she runs towards her mother with her arms outstretched. She jumps into the open arms of her mother, who is laughing and crying at the same time. Her father, a huge man, completely engulfs his wife and daughter in a giant but gentle bear hug, while breathing out a loud relieving "phew."

After a few moments, the parents look up and begin to eye me quizzically. Oh dear, I think.

"I know you."

"Do you?" say I, not recognizing them.

"Yes," says the mother, "you were our family doctor in the '80s. You must remember looking after nana while she was dying."

She tells me her nana's name and, indeed, I do remember, picturing the face of a brave and dignified lady in her late sixties who died at her home of bowel cancer.

I am then ushered into a four-way hug, father naturally on the outer rim, while I try to explain my predicament that before had seemed so worrying but now seems so trivial and unimportant. They listen politely and patiently. But I realize, at this moment, for this young family only one thing matters. They are simply together again.

In the following pages, I will explain to the best of my ability why the holographic paradigm has growing scientific validity. But ultimately, my reason for exploring and explaining this whole subject is purely practical: if we are indeed on the threshold of confirming that realities exist beyond the boundaries of our four dimensions, and we are indeed holographic beings vitally and instantly interacting with a holographic universe, what does all this mean to our everyday lives? Does it enhance and enlighten us, or merely satisfy our thirst for knowledge of the unusual? If we all indeed manifest from a universal field of consciousness, that consciousness and pure information is our primary source, so how does the knowledge of this affect and alter our behavior?

Every week, perhaps every day, each of us faces the type of dilemma I encountered at the beach. My story is intentionally rather ordinary to illustrate that we do not have to rescue, resuscitate, or perform lifesaving surgery with a penknife to play our part. The daily challenges we face may not seem hugely significant to the outsider, but by meeting them and by acting wisely, we contribute to the gradual expansion of our own consciousness and, in turn, to the collective pool of universal consciousness. At the beach on a summer's day, while floating blissfully in the calm warm water, it is easy to imagine that we are part of a perfectly benign and harmonious field that encompasses all that is wonderful about our world. (See figure 2.)

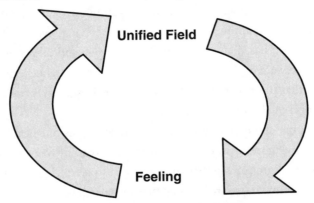

Unified Field

Feeling

Figure 2. At one with the field.

In fact, for years I have encouraged patients to visualize, even feel, the sensations I described in my beach story in an attempt to guide their bodies into a perfect healing state. Because many of us live our lives at a high pace, under strain and in a rush, this exercise is indeed a fitting antidote to the many stresses of modern-day living. Occasionally, someone leaves my rooms in just such a state of deep relaxation.

Acupuncture, reiki, or massage can all lead to the recipient feeling this way, and it can be truly life-altering for someone unaccustomed to achieving a level of bliss using purely natural methods. I have witnessed those previously addicted to the effects of drugs, alcohol, and cigarettes change their habits dramatically once they discover that a state of deep happiness can be reached in an instant solely through activating their own inner resources.

Others become aware through peak experiences that occur out of the blue in their everyday lives. Every golfer experiences an instant in each round when everything is just right—the ball flies perfectly, and no one, not even Tiger Woods, could do better. Unfortunately, in my own case, these precious moments are all too fleeting. It is precisely the memory of these rare occurrences, however, that brings us all back to the golf course week after week. Much of modern sports psychology is dedicated to enhancing and building on these peak "in the zone" experiences through visualization exercises.

Tiger Woods, according to his Kiwi caddie, Steve Williams, visualizes before each drive the precise spot on the fairway where his ball will land, and also where it will end up. Of course, for this to work, he also has to have the discipline and will to practice for many long and lonely hours. And it helps to have Tiger's truly unique innate talent. So achieving, even wallowing in, a dreamlike state has many wonderful advantages, and allows us to stand outside the pressure of time that so depletes us of our vital energy. These are experiences that place a true perspective on our lives, away from the purely material world of "I win" and "you lose."

Meditating alone or in groups not only enhances our health, but also leads us into living more productive and creative lives. By focusing our intent in a coordinated group, whether in prayer or as part of Lynne McTaggart's Intention Experiment, we are learning that we have the power to produce life-enhancing effects on others nonlocally. And by connecting instantly to the now through energy psychology exercises, we are learning to delete destructive conditioned and addictive behavior patterns quickly and completely. But learning how to live our lives in harmony with the unified field, the subtitle of this book, however valuable, is only one lesson the holographic paradigm teaches us.

As I lay in the warm water that late summer's afternoon, I was filled with a sense of awe at the perfection of our world. I felt a sense of deep gratitude for this. But the uncomfortable events that were to unfold a few minutes later on dry land were, in retrospect, more profound. The tangible world—of fear, time pressure, worry, and guilt—I then encountered set the scene for me to try my best to do the right thing. The relief I felt at the joyous outcome—a child reunited with her parents—reminded me, yet again, to trust in a process that unfolds once one's intent is set. (See figure 3.)

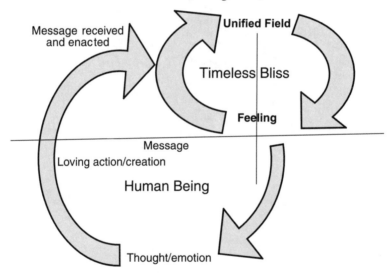

Figure 3. Feedback with the unified field.

We have all been granted the free will to act appropriately, to participate in our world, for the greater good. The free will to swallow our pride, to feel uncertain and fearful, to admit that at times we have been wrong, and, above all, to be kind to ourselves and others. In the holographic universe, one part, no matter how small, can influence the whole. An act of kindness from the heart of a humble slave matches that from the heart of a sovereign.

First Principle of the Human Hologram: The study of holographic science and philosophy gives us a deeper understanding of the value of our everyday lives.

Section Two
The Human Hologram—The Science

Chapter 1: Introduction to the Science

All things by immortal power, near and far
Hiddenly to each other linked are,
That thou canst not stir a flower
Without troubling of a star.
—Francis Thompson (1859–1907), British poet

The holographic paradigm wherein all parts of a greater universe are expressed fractally in each smaller part has been intuited by visionaries, artists, and poets for millennia. Indeed, as Karl Pribram summarized in his 1991 book *Brain and Perception,* the brain, senses, and mind of a human being may be inextricably involved in the process of relaying this universal information. Furthermore, there is the growing evidence that our bodies are continually being formed upon a holographic matrix "at one" with a universal field of consciousness.

But in the world of orthodox science, heavily based as it is on mechanistic Cartesian principles, such metaphysical philosophies remain difficult to prove. Even though we have known for over a century that time is relative, that matter and energy are interchangeable, and that particles can just as accurately be regarded as wave forms, scientists are still required to base their theories on the laws of the universe that can often in the academic world be seen to lag behind the insight of visionaries.

It also appears a great struggle for many, in their professional lives at least, to stand outside themselves and entertain the concept that their own observations, their own perceptions, and their own being can affect the object they are studying. The cosmos is still perceived of as being "out there" in the distance, completely separate from us, the observers. But the revolutionary information age is rapidly changing the ways ideas are being disseminated to the general public. Recently, I gave a talk at a sold-out conference for which no money had been spent on advertising. The registration details had been spread exclusively "virally" over the Internet through contacts on sites such as Twitter and Facebook in a web-like fashion.

Ideas can now reach many people trained in many disciplines who share enthusiasm and vision. Intelligent enthusiastic "amateurs," not threatened by thoughts of damaging their reputations or losing research funding, now discuss new theories alongside established academics and research scientists who are secure yet still inquisitive enough to push the boundaries of knowledge. Fears that this liaison could prove dangerous and contribute to a dumbing-down of our knowledge base appear, by and large, to be unfounded. And as a result of this new explosion of information free to the masses, education programs are being developed based less on cramming facts into students and more on ways they can access information that carries value and integrity.

As a family doctor, I have had the opportunity to study and practice medicine based on holistic and holographic principles. The Law of Five Elements in Chinese medicine allows us to view the workings of the body alongside the laws of nature—the macrocosm of nature expressed within the microcosm of the body. This perennial wisdom has helped me greatly to gain a deeper understanding of health and the healing process. I have found that this knowledge can be easily passed on to the people seeking my help and can help them manage their health conditions in real, practical ways.

Choosing, and using, nature to help heal—both through the food we eat and through the environment in which we live—is more important than ever in the fast-paced twenty-first century. Similarly, understanding the principle of yin and yang leading to achieving balance in our modern lives is as essential now as it has ever been. In fact, this basic law within Chinese philosophy is precisely the same as a basic law of Western science: the *first law of thermodynamics.*

In its simplest form, this states that energy in this universe cannot be destroyed, merely converted to other forms. So, fundamentally in nature, there is always an underlying tendency towards a state of perfect balance, or symmetry. Nothing in nature is wasted. Leaves fall from trees to the soil, which is enriched and fertilized by their decay. Farmyard manure, or better still, "zoo poo," makes the tomatoes from my modest vegetable patch taste, according to family members, "almost like bought ones." Water rises from the seas to condense as clouds, which eventually release over the land to irrigate and support all earthbound life. In modern terms, one could say that it is the pure basic "information" held by all these differing forms that remains intact. There is now a growing understanding that this universal information is simply another way of explaining a universal consciousness that is fundamental to all we know.

But life on this earth brings us all face-to-face with another, more inescapable, truth. Three and a half decades of Western medical practice, and nearly six decades on this planet, has brought me face-to-face with the irrefutable fact that our bodies sadly deteriorate with age, requiring more attention and self-care as the years pass. So in the third- and fourth-dimensional realms of this earthly life, we see *entropy* in action precisely as expressed in the *Second Law of Thermodynamics*. This shows that a complex system such as the human body progresses in time irreversibly towards a state of chaos and disorder that is more and more difficult to "fix." In fact, without this display of entropy, as a doctor I would probably be out of a job! All complex machines are subject to entropy too; hence the perpetual need to back up data in fear of our computer hardware crashing. It is a reality for all of nature, as even the humblest amoeba has an information storage system well ahead of the most powerful man-made computer. If I drop an uncooked egg on the floor, it will inevitably smash and, as with the poor unfortunate Humpty, "all the king's horses and all the kings men" (even IT repairmen!) will be unable to reassemble my breakfast.

I recall a lovely Buddhist nun being interviewed on the radio, explaining how we shouldn't grieve or be annoyed when a precious heirloom such as a vase smashes on the floor, as it is simply going the way of all nature. In fact, without these events, antiques that remain would fail to become valuable.

The insights of the twentieth-century physicist David Bohm lend a deeper understanding to the two laws of thermodynamics. To describe our three- and four- dimensional world of entropy and decay, he coined the term "the explicate order." And the symmetrical, balanced realm that transcends time and space he called "the implicate order." There is a growing contemporary understanding that our explicate time-stressed world is a virtual reality concealing many a timeless implicate spiritual truth.

The times when the veil between these worlds becomes lifted appear to be at the extreme ends of our lives, or at times of deep emotional meaning. It is then that such truths are known as clearly as we perceive that grass is green and sky is blue. Many who live through near-death experiences gain insights into eternal truths that place the superficial pleasures we seek from our material world back into their true context.

Chinese and Taoist philosophy can also help us understand how, within our explicate world, we can create perfect balance between positive (yang) and negative (yin) forces, and access the limitless realms of the implicate order, variously referred to in science as zero point, scalar, or free energy. The middle way is often the most fulfilling in our lives—for example, when our masculine and feminine energies balance, when there is win-win rather than win-lose. Yet still, our media is more interested in selling us the unusual, the bizarre, and the extreme.

Symmetrical geometric shapes, the symbols of sacred geometry, occur fractally in the natural world as ever-persistent signs to us, providing us with evidence of dimensions beyond the purely physical—symmetry beyond the chaos. For example, Fibonacci series, and the related Golden Mean of 1.618, are observable everywhere throughout our cosmos from within the spiraling galaxies observed though our most powerful telescopes to the structure of our DNA accessed through our cutting-edge electron microscopes. They are consistent in the growth patterns of shells, vegetables, and cells in the human body. Sacred geometry strongly hints that all living beings, our planets included, are universally connected, and that separation may just be an illusion. We will explore fractals in greater detail, including their expression of complexity, in chapter 9.

My exposure to Ayurvedic medicine has helped me apply these unfolding patterns to our growth of consciousness, and enabled me to give some meaning to the difficulties of those seeking my help. Our life's journey on this earth involves many real-life encounters in which a balance has to be achieved between acceptance and action. This has helped validate to me the concept of chakras and kundalini, our inner spiral staircase to a place of perfect peace. The chakras also connect us holographically to the whole, so our individual journey feeds directly and instantly into the whole community. Therefore, as our life unfolds, so too does the great unifying field of consciousness.

Applying this model, each thought, each observation, contributes to the cosmos we are observing. The things we observe with our senses are merely our perceptions of the world around us. Conditioned restrictive perceptions fed to us by others, especially in childhood, may have to be reexamined and even replaced by new truths. It has recently been discovered through functional MRI (fMRI) brain scans that light reaching our eyes is processed not only by the so-called visual cortex of our brains, but widely through other cortical areas as well. So it is more than possible that conditioned beliefs, and narrow perceptions, influence what we think we actually see. Beauty, it is said, has always been in the eye of the beholder.

In the third section of the book, we will explore how the deep healing process involves being guided and supported through these changing perceptions. The healing journey follows the precise fractal pattern of a life lived from early childhood to independent adulthood. As an infant and young child, we feel intensely but lack the brainpower, the rationality, or the eloquence to put these feelings into words. The feelings move us into action via the expression of our emotions. It is then that our brains, and our thoughts, become important. If there has been imbalance in our childhood, that is, feelings not acknowledged, our emotions stifled, then suppressed feelings can submerge into our subconscious. This can result in disease in later life.[3, 4] So according to this model, heart-based feelings can be said to be more fundamental than brain-based rational thoughts. We need to employ our hearts and our heads in a balanced way if we are to heal ourselves and our planet. A society based solely on purely rational decisions devoid of compassion is a dangerous, and an endangered, society.

It is doubtful that my own interest in holograms would have either been sparked or sustained had I not decided to stick a solitary pin into the auricle (the outer part) of someone's ear, one morning in the early 1980s. The fact that this bizarre therapeutic intervention could somehow prove to be anything more than a brief distraction from my truck-driver patient's low back pain seemed then to be far-fetched in the extreme. Yet, despite this, I did it and the truck-driver was instantly relieved. And I have repeated this simplest of procedures every week, almost every day, of my practicing life to date.

This year, a young engineering student in the midst of a severe migraine attack came to the university health center. His usual painkilling medication, and an anti-inflammatory injection, hadn't worked, and he was due to sit his final exams one hour later. He didn't want anything that would make him drowsy. Sportingly, he agreed to an acupuncture needle in his ear, which gave him instant relief, and remained in place for the duration of his exam, much to the amusement of the exam adjudicator. Necessity, an engineer would be the first to remind us, is so often the mother of invention.

Well-designed randomized controlled trials are now being conducted to investigate the efficacy of ear acupuncture. In 2007, following a pilot study, doctors at Ernst Moritz Arndt University in Germany showed that tiny needles inserted into ear acupuncture points reduced pain after knee surgery. The needles were taped in place for the duration of the operation; the control group of patients had needles inserted into places in their ears that did not correspond to known acupuncture points. The patients receiving the correct acupuncture required significantly less pain-relieving medication after their surgery. The paper, published in the *Canadian Medical Association Journal,* provoked much debate. One skeptical doctor was highly critical that a reputable peer-reviewed journal should publish such a study, calling the trial "a waste of money."[5]

The knowledge *that* a medical intervention actually works often precedes an understanding of exactly *why* it works. In 1746, Scottish naval surgeon James Lind began to study why so many sailors in the British Navy were dying of scurvy during long voyages. Through diligent research, in what would prove a beautifully designed controlled trial, he showed that sailors on the *HMS Salisbury* who included lemons and oranges in their diets were able to reverse completely the signs of this potentially deadly disease. Unfortunately, it was to take a further forty years, and 100,000 deaths, before authorities formally introduced lemons and limes into sailors' diets. And it was not until 1932 that Hungarian scientist Albers Szent-Gyorgyi and English chemist Walter Haworth finally isolated the chemical responsible, vitamin C, naming it ascorbic acid after the Latin word for scurvy, *scorbutus.*

Second Principle of the Human Hologram: Holographic theory builds on ancient wisdom and modern science at a critical time in our history when information is being shared freely around the world, breaking down old barriers of race, religion, and politics.

A Brief History of the Hologram

The hologram was pioneered by Hungarian scientist Dennis Gabor in 1947. Gabor had spent his early working life in Germany, but fled to England from the Nazi regime in 1933. His groundbreaking work remained theoretical until the invention in 1960 of the laser as a source of coherent light. Coherent light waves are those that are formally "in step" with each other. Over the next few years, scientists in Russia, the United States, and Britain subsequently developed prototypes of the hologram.

Two beams of laser light were produced from one source. One was aimed at the object to be represented in the hologram, while the other was diverted to a mirror. The light from the mirror, known as the reference beam, was then reunited with the light that had been scattered, or diffracted, by the object. This resulted in a complex interference pattern of light waves, and it was this pattern that was recorded on a special photographic plate. (See figure 4.)

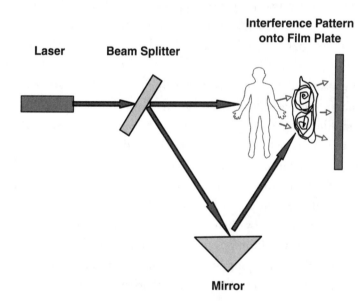

Figure 4. Recording the hologram.

When this flat two-dimensional photographic plate holding the information was exposed again to a fresh laser beam, a ghost-like three-dimensional image of the object appeared, reconstructed from the information stored on the photographic plate. So all the information was carried within the beam of laser light forming the holographic image. (See figure 5.)

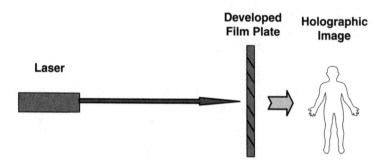

Figure 5. Developing the hologram.

Since the hologram stores and then reconstructs all the scattered light from an object, the image is seen by the viewing human as being as real as the original object. Two people on opposite sides of the hologram will see different sides of the holographic image, just as if it were real. When we move around the hologram, the object stays still and we can explore it from different angles. This property is known as *real parallax.* This is not so when we don our 3D glasses at the cinema, as everybody in the audience sees exactly the same image, whether they are sitting in row A seat 1, or row W seat 20!

The other important property of the hologram, and essential to this book, is the fractal nature of the image, that is, that the smallest part of the recorded information on the film contains, when exposed to a single laser beam, all the information of the whole. This can be demonstrated by first cutting a small fragment out of the plate onto which the holographic image has been recorded, and by then subjecting it to a laser beam. A complete holographic image will form as a result. (See figure 6.)

Just why the overlapping beams combine to encode the information in this startling way is still the subject of intense debate. When two sets of ripples on a pond converge, some will overlap, doubling their height, while others will cancel each other out—a process known, respectively, as constructive and destructive interference. A variable mixture of these patterns occurs as the two laser beams meet—one carrying the information of the object and the other acting as a reference. In recent experiments, interacting laser beams have been shown to provide the perfect environment to induce entanglement, or nonlocal instant connections between atoms.[6]

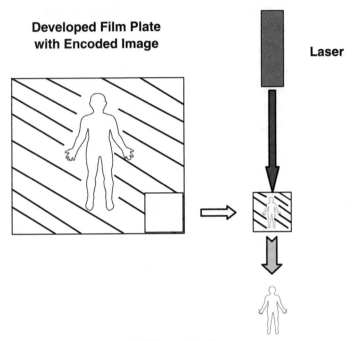

Figure 6. Holographic fragments.

As the light waves interfere and cancel each other, a medium or vacuum results, open to information beyond our known confines of time and space. Subatomic particles trapped in these converging "headlights" respond instantly by oscillating and linking up in entangled pairs with opposing spins. A perfect harmonic resonance thus forms between light, atoms, and their entangled twin atoms. This concept will become clearer as we explore in chapter 5 the fascinating role of entanglement.

So the hologram contains both local and nonlocal information, carried within a beam of laser light, with input from several sources:

1. The shape of the object via the diffracted beam of photons (local)
2. The relative position in space of the object via the interaction with the reference beam (local)
3. Nonlocal "universal" information from the "destructive interference" of the beams of photonic light.
4. Our own observation (We'll explore this in chapter 3.)

If, indeed, we extrapolate holographic theory to its logical conclusion, all information of the whole can be conveyed by the smallest part. So each photon could itself be carrying, or at least transmitting, universal information. This property is demonstrated in the final step of Gabor's hologram experiment, as the fresh beam of laser light "absorbs" and "records" the information stored on the plate before forming the holographic image. No one has actually seen a photon, but there are those, including artist Jon-Henrik Andersen[7] and a team of physicists at the University of Michigan, who depict the photon as a toroidal or donut shape—the shape of a wormhole into which all structures from this dimension eventually dissolve and, quite possibly, out of which all structures continuously evolve. We are now applying the new paradigm of holographic theory and touching on truly mind-stretching concepts!

Dennis Gabor's pioneering work was formally recognized in 1971 when he was awarded the Nobel Prize in physics. Since then, holograms have become an integral part of modern daily life. As they are very difficult to replicate, requiring the use of expensive equipment, they are to be seen as security on many banknotes around the world. As their information storage potential is huge, they are being touted as the next significant advance in data storage. Beyond this, they are used in fields as diverse as supermarket scanning and electron microscopy.

But since the 1970s, leading physicists have been exploring the possibility that our very universe is, in its most fundamental form, a hologram. To understand this, please join me as we enter into the dark, rather uninviting abyss of a black hole.

Third Principle of the Human Hologram: The man-made hologram has two prominent properties. First, that of real parallax in which the image appears to remain in a fixed position when viewed from different angles. And second, it has a fractal nature whereby all parts of the whole are contained in the smallest part.

Chapter 2:
The Universe as a Hologram

In the 1970s, Stephen Hawking showed that black holes, into which it appears stars and galaxies (and ourselves, if we accidentally strayed there) inevitably dissolve and "evaporate," are not entirely black. In fact, he showed that they do slowly emit radiation, causing them eventually to disappear entirely. The laws of entropy insist, however, that energy cannot be destroyed in our universe, only recycled. When something decays, it has to decay into something else—as in "dust to dust, ashes to ashes." So black holes seemed to provide an exception to this rule, as everything supposedly disappeared into nothing. This has been termed the black hole paradox.

A solution to this was offered by Jacob Bekenstein of the Hebrew University of Jerusalem in Israel.[8] Surrounding every black hole, there is, in theory, a flat curved border, known as an event horizon (see figure 7). Bekenstein, through complex mathematics, proved that all the information within the vast volume of a black hole can actually be held on its flat "event horizon" surface. For this to happen, obviously each storage bit on the surface has to be black-hole infinitesimally small, in fact, the very smallest measurement known to science: Planck's constant. But strangely enough, the math does work without—and this is important to scientists—breaking any of the current laws of physics. So the energy of something entering a black hole doesn't have to completely disappear into nothingness; it is merely converted, and encoded, into some of the trillions upon trillions of Planck's constant-sized pixels on the vast surrounding mega-screen.

Bekenstein has since received support from other pioneering physicists, most notably, Leonard Susskind and Nobel Prize–winner Gerard 't Hooft, who have extended this theory, applying it to our universe as a whole. This is truly remarkable when one realizes that, according to our current understanding, our universe is pretzel-shaped, "encased" within an event horizon nearly fourteen billion light years away. Again, almost incredibly,

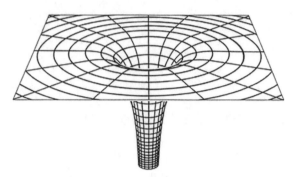

Figure 7. Black hole—event horizon.

the math works if one considers that this enormous two-dimensional screen comprises trillion upon trillion of information-holding dots, each one the size of Planck's constant.

If this were the case, then our time-space reality could then be envisaged, before honed by our human senses, as a magnified blurred version of all this information projected into "thin air," like an out-of-focus laser light show. And bizarrely, this is exactly what a sophisticated cosmic energy measuring machine in Hanover, Germany, could be showing. The GEO600 is a highly sensitive gravitational wave detector stretching six hundred meters that has, since 2002, searched for subtle waves of energy emitted from distant dense stars and black holes. To date, it has yet to detect the waves. Instead it has detected a fuzzy, grainy interference, entirely consistent with Beckenstein's theory and calculations. In fact, he had predicted in advance that the cosmic projections from the surface of our universe would cause the precise patterns that are now being found.[9] At the time of writing, Craig Hogan and his team of physicists at Fermilab near Chicago are designing a 40-meter-long "holometer" to test Beckenstein's theory, in an attempt to confirm and measure this fuzzy "interference" more precisely. If, indeed, this and other research supports this theory, it could lead us to the inevitable conclusion that it is our eyes, ears, and touch that make sense of this information, converting it all into a tangible reality, a medium through which we can achieve our goals here on Earth.

This process has parallels in our everyday life. Our flat television screen contains thousands of tiny dots that continuously change color. Photons or waves of light emitted from these dots reach our eyes, from whence they mix, mingle, and cross over a number of times on their way to our brain where they blend with our preconceived perceptions. The TV speakers emit energy, vibrations that travel through the air of our living room and reach our eardrums, causing them to vibrate in turn. Again, our brain processes this energy, turning it into what we know as sound.

This combination of light and sound draws us into a virtual reality that can move, entertain, and exasperate us. Our sense of touch and time-space awareness helps convince most of us that TV land is, in fact, an illusion. We can tap the screen, look behind the set, and convince ourselves that Larry King hasn't joined us for the evening. However, with the advent of 3D television technology and holographic projection, discerning just what is and what isn't real may become more tricky. I imagine our poodle, Lily, could become very vocal if a meowing 3D cat was to appear suddenly on the coffee table. But for the rest of us, this technology is leading us towards a closer understanding of how we perceive reality—or what we have always regarded as reality. The virtual reality of computer games both entertains us and intrigues us. It is also a phenomenon mirroring the growth in our conscious awareness, drawing us towards a deeper understanding of hidden realms that lie behind and beyond our material world, encoded at the very margins of our reality. We are beginning to realize that all we thought of as being real, including our physical bodies, may be contrived perceptions whose existence is dependent on our senses.

These are, of course, speculations. But they are based on the conclusions of a growing number of cutting-edge twenty-first-century seminal thinkers—physicists and mathematicians with staggering mental capacities and powers of deduction. In the current harsh academic environment, however, their hard-earned reputations could be put in jeopardy were they to be seen to expand recklessly on these theories. But unencumbered by fears of derision by our peers, you and I are perfectly free to speculate further.

As the physicist Robert Dicke once said: "The right order of ideas may not be, 'Here is the universe, so what must man be?' but instead 'Here is man, so what must the universe be?'" Each of us must now try to understand our own special role in this miraculous process of cocreation, using a universal gift common to all sentient beings, irrespective of their IQ level, or even their number of legs (or wings): our powerful gift of observation.

Fourth Principle of the Human Hologram: Modern scientific and mathematical theory lends support to the theory that our universe is holographic. It follows that we, as part of the universe, must also be holographic. Modern science acknowledges that pure information is fundamental to our universe.

Chapter 3: You the Observer

Who are you going to believe, me or your own eyes?
—Groucho Marx (1890–1977), American comedian

The perception that we are separated from others, our world, and our cosmos has not been without its benefits. As young adolescents, we need to identify our points of difference, the physical and personal strengths that will best allow us to serve our families and communities. First, our egos need to be formed, then to be balanced with humility and empathy—the deep appreciation of others. As a result of this personal, spiritual journey, our own lives, and the lives of others, flourish as we bring our unique presence, our being, to the world. Our wisdom evolves and grows.

But our separateness from our environment is nevertheless, at the very essence of our being, an illusion. In reality, the observer and the observed are engaged in an intimate dance. Or as the twentieth-century philosopher Bertrand Russell explained: "The observer, when he seems to himself to be observing a stone, is really, if physics is to be believed, observing the effects of the stone upon himself." The subject and the object are connected. Photons from the sun are diffracted from the surface of the stone, to meet each of our eyes from a slightly different angle. As with the photons from our TV screens, they are then converted to electrical impulses at the back of our eyes, which then pass to different areas of our brain for processing. And even without touching it, our past experiences will tell us whether it is likely to be something we could pick up with one hand, or whether some poisonous creature lies hidden underneath.

Some things we know exist only in our minds. When we hear scientists describe the real universe as being encoded on a flat screen—strange though this concept is—we can, indeed, imagine this. We may recall TV or movie screens, and possibly even images of balloons and pretzels.

If we were able to peer into one of the countless number of atoms that combine to make our bodies, we would find tiny particles within vast expanses of space. We can picture this, as it conjures up an image of a night sky with its tiny stars surrounded by similarly vast areas of nothingness. Our perceptions are dependent on our experiences.

But quantum theory takes our understanding of the role of the observer to yet another level. Experiments repeated time and again over forty years show that electrons that circulate the nuclei of atoms only collapse down into particles when we influence them through the attention of our minds. Before this, physics tells us, they exist only as undetermined states, or waves, of probability.

Similarly photons behave either as individual packages of light, or wave forms, depending on how we want to view them. Before this, modern theory goes, they too exist only in realms of probability. We can view light as either consisting of individual packages known as photons, or in wave form; but we cannot perceive light as being both at the same time. The term for this is *complementarity*. The famous, oft repeated and quoted, twin slit experiment demonstrates this conclusively. The wave nature of light, challenging Sir Isaac Newton's particle (or *corpuscular*) theory, was first demonstrated in 1802 by doctor, physicist, and Egyptian hieroglyphics scholar Thomas Young.

Young observed how the ripples on a pond interacted when two stones were thrown, breaking the surface of the water side by side. He wondered whether this could explain the strange striped pattern that showed on a screen when light shone through a solid sheet into which two slits had been cut side by side (see figure 8a). He compared this with the pattern produced when light was shone through a single slit (see figure 8b).

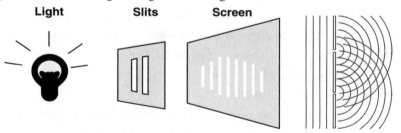

Figure 8a. Double slit–interference pattern. The pattern on the screen represents columns of overlapping waves. Like ripples on a pond (right), if two waves meet, they will either interfere by adding to each other (constructively) or by canceling each other out (destructively). The pattern represents multiple waves interacting in this way.

Light   **Slit**   **Screen**

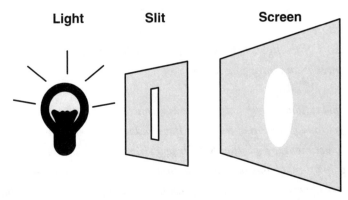

Figure 8b. Single slit–no interference pattern. Light through single slit. Light spreads out in a logical fashion. To demonstrate this to yourself, try shining a light through a slit in a sheet of cardboard. Alternatively, imagine how pellets fired from a shotgun spread over their target.

For the past two hundred years, scientists have continued to explore and expand on the many mysteries of Young's findings, employing ever-more sophisticated methods and equipment. Modern advances in laser technology and sophisticated detectors are now used to refine the light source and to record the results accurately, allowing us to test his theories in greater detail. For example, we are now able to set up an experiment that alters the light in such a way that we can identify exactly which of the two slits it goes through. When we do this, the light appears on the screen, not in the striped pattern of interacting waves, but in the pattern of individual "bullet" particles. Figure 9 illustrates this in more detail.

Laser   **QWP**   **Slits**   **Detector/Screen**
Light

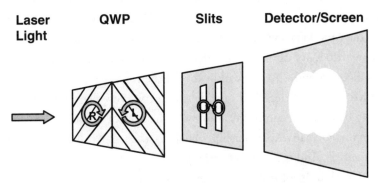

Figure 9. Changing (circular) polarization with quarter wave plate (QWP). Observer can predict.

This experiment uses a coherent laser light and a plate known as a Quarter Wave Plate set in front of the slits, and a detector plate, which records the position and polarity of the light. For simplicity's sake, I have illustrated the outcome (no interference) as an image on a screen.

Explanation: If light was to exist as a collection of bullet particles, it would have to go through one slit or the other. If we had a way of separating light into two (in this experiment, by splitting circular polarities), we could envision one band of bullets going through one hole and one through the other. Polarity-altering plates (known as Quarter Wave Plates) are used to fix the polarity of the light, which is passed through each of the two slits and then recorded on a detecting plate. So, by recording the polarity, we get to know exactly which slit the light has passed through. If we repeat the experiment armed with this which-way knowledge, the pattern appears as a particle pattern—not as a wave-form pattern. This happens with different types of detectors and splitters, with the evolving consensus that this is a direct "observer" effect. (See appendix 2a for more details, including an explanation of polarity.)

So it appears that when we, as observers, expect light to behave as a collection of individual bullets, it responds by doing precisely this. If we expect it to behave like waves, it so obliges.

In the past ten years, the strangeness of this interaction at the smallest, most fundamental level of being has become stranger still. It has recently been shown, using entangled photons, that our influence on this quantum world can occur outside known time lines. Observation after the event can affect the process at source—retroactively. Also attention and intention from a distance does not diminish this effect; it is truly nonlocal in character.[10] (See appendices 2a and 2b.) So, within these experiments at least, we cannot be regarded as passive detached outside observers. Our own minds and intent seem to be in direct communication with the subject (light) we are studying, forming a cocreative partnership in which each party has some, but not total control. And it seems our minds, when engaged, can cut through our conditioned limits of time and space.

And as we are clearly intimately part of this process, this science must shed light on our own timeless ethereal nature. Somehow, in the past century, our science has been able to penetrate through the restrictions of our time-space to encounter other dimensions, previously the domain of mystics and shamans. At the heart of our sophisticated twenty-first-century rationality lies a seed of uncertainty, or wonder, that, if nurtured, will grow and show us more of itself. Our happiness is so often linked to how we handle, and welcome, uncertainty. Confidence in our ability to change and adapt frees us to live our lives in the present moment.

The sixteenth-century philosopher and founder of the scientific method Sir Francis Bacon once remarked: "If one will begin with certainties, one will end in doubts, however if one were content to begin with doubts, one will end in certainties."

Fifth Principle of the Human Hologram: The science of the human hologram must include an awareness and study of our own participatory role as observers of ourselves, our world, and our universe. In *observer physics,* we become aware there is an ever-present subtle partnership between light, matter, and ourselves.

Chapter 4:
The Cat, the Moon, and the Parking Space

In the quantum world, a subatomic particle exists only in a state, or wave, of probability before our observation allows it to manifest itself. Famously, in 1935, Austrian physicist Ervin Schrödinger questioned whether this rule applied to the macro-world of living creatures—for example, his cat. Would a cat trapped in a shielded box and unfortunately poisoned by cyanide from a vial shattered by a random radioactive process be, prior to discovery, alive, dead, or in an undetermined preexisting state? Or indeed, as Albert Einstein reputedly asked physicist Neils Bohr, does the moon really only exist if someone is looking at it?

To date, experiments have yet to prove such an observer effect (by individuals) on such a large scale, although Oriol Romero-Isart at the Max Planck Institute in Garching, Germany, is presently devising a study involving a "Schrödinger's virus" in the place of a cat.[11] The popular movie *The Secret* suggested this could work on an everyday level—using the power of our attention to manifest central-city parking spaces. Disappointingly, this has never worked too well for me.

Recently, I was asked, along with a number of others, to a group audition for one of the many movies being made in the wake of this hugely successful film. We were to meet with the producers in an office block in the heart of downtown Auckland, a city poorly served by public transport. I arrived in my car twenty minutes early, and along with several other candidates, began circling the surrounding streets searching in vain for a parking space. Half an hour later, I returned home, resigned to the fact that this was one gig I was clearly not meant to attend. I still wonder, however, whether this was, in fact, a clever and cunning plan devised by the producers to weed out those still unable to create effectively their own reality. A test of transcendence that, alas, I failed to pass.

Later, we will explore the concept that it is compassion and togetherness rather than competition and separateness that facilitates such quantum effects in our day-to-day lives.

Sixth Principle of the Human Hologram: The observer effects we encounter at the micro quantum level do not always help us find that perfect macro parking space.

Chapter 5:
Entanglement—All Together Now

"An instant, an aspect of nature contains all of nature." These translated words written large above the entrance to the 2009 Claude Monet exhibition at Wellington's Te Papa museum were those of the master French impressionist himself. For the last thirty years of his life, Monet was at his happiest painting everyday scenes of nature in his two-acre garden at Giverny. Every single moment, every subtle change of light presented Monet with a different subject to capture on canvas. He rarely felt the need to travel away from home to observe and record the beauty of his world, preferring instead to explore, and revel in, the microcosm that surrounded him at his home. One painting, *Le Bassin Aux Nympheas,* from perhaps his most famous series, *The Water Lilies,* sold in June 2008 at auction in London for a record US$80 million.

To really understand "entanglement," one need look no further than at a work of art such as *Le Bassin Aux Nympheas*—a timeless record of one man's instant impression of nature. Every part, every completed brushstroke complements the whole. A world on record with no obvious linear cause and effect, and with no complicated equations or extrapolations. Over the past century, science has moved closer to understanding just how our world, and our universe, works. Physicists studying the tiny quantum realms have had to face and resolve many challenges to their conditioned views of nature.

The old rules of science separating us from our environment and regarding time and space as fixed no longer seem tenable. In 1935, the trio of physicists Albert Einstein, Boris Podolsky, and Nathan Rosen humbly came to the conclusion that the odd results they were getting were due to "hidden variables," properties of matter and energy that they as yet didn't understand. Einstein struggled with the concept of nonlocality (instant telepathy between particles or "spooky action at a distance"), explaining it away as something independent of our observation not yet fully understood.

At his death in 1955, Einstein still hadn't reconciled with what has come to be known as the Copenhagen interpretation of Heisenberg's uncertainty principle, which holds that our observation or measurement is an integral part of quantum physics and, in turn, our "reality," and that subatomic particles can exist as mere probabilities before materializing. Einstein found it difficult to believe that God would allow such unpredictability and randomness, "playing dice" in this manner. As with other scientists before him, his discipline required him to distance himself from the outcome of any experiment. It was not until ten years after Einstein's death that Irish scientist John Bell showed through mathematics that the true reality of this small quantum world was indeed one of instant, nonlocal connections, and that our measurements themselves were statistically involved in the outcomes.

Largely as a result of this work, known as Bell's theorem, we are now aware that whenever we look for an electron spinning around the nucleus of an atom, it appears out of the blue. However, it spins precisely the way it wants (our looking doesn't have total control), with its entangled mate instantly spinning the opposite way (see figure 28). The mate doesn't have to be in the same room, town, country, or even galaxy. There is no cause and effect here; this, as with Monet's *Water Lilies* at Giverny, is simply a state of togetherness or coexistence.

In 1997, Nicolas Gisin and his team from the University of Geneva sent entangled pairs of photons through fiber-optic cables to detectors ten kilometers apart. They found that measurement of one photon instantaneously influenced the result of the other. This was strong experimental proof of nonlocality—evidence of a connectivity within nature beyond our concepts of time and space, and separateness. And as mentioned previously, further experiments on entangled pairs of photons in modified versions of the twin slit experiment this century have confirmed not only that photons can connect nonlocally, but also that human attention, and possibly even *intention,* can affect results *retroactively,* breaking through our perceived barriers of time and space. But how easy is it to study entanglement between other subatomic particles apart from photons? How about electrons, whole atoms, molecules, and organisms?

Until now, to demonstrate entanglement between electrons in atoms, scientists have had to create conditions at the very margins of our material world. For this to happen, the environment has to be cooled (via lasers) to temperatures just above absolute zero—the very limit of coldness below which life is unsustainable. At this temperature, not only is it possible to observe entanglement between electrons and atoms, but also (at present, only momentarily) between tightly packed molecules the shape of modern soccer balls, geodesic spheres—the so-called Buckyballs.* So there is excitement around this finding in the information technology (IT) and telecommunication industries, as it represents faster and more efficient ways to store and transport information. Since information of a Buckyball's structure is instantly shared, the potential exists to clone or teleport such tiny objects.

And there is even talk within mainstream science of teleporting living things. There is an eight-legged microscopic creature known affectionately as a water bear, and more officially, as a tardigrade, that has been thriving on our planet for half a billion years. It is an amazingly resilient animal. It can be found at the bottom of our deepest darkest seas, and on the top of our highest snowcapped mountains. It can survive a decade without water, and tolerate extreme heat and high levels of radiation. Even more fascinating, it can radically reduce its metabolism to a near-death-like state in a reversible process known as *cryptobiosis*. In September 2007, it was studied in the vacuum of outer space on the Biopan-6 experimental platform provided by the European Space Agency during the FOTON-M3 mission. Not only did the water bears happily survive their experience as astronauts, they continued to reproduce in the extremely dehydrated vacuum of space. What is more, they are known to survive temperatures approaching absolute zero, hence are being considered as ideal experimental miniscule "guinea pigs" for quantum cloning.[12]

At this low temperature, everything slows down. Lasers are used to chill atoms to this state of "suspended animation." Marcus Chown, the British science writer, explains that this happens because the laser beams' photons absorb the energy of the subatomic particles, bringing everything to a halt. In this state, atoms lose their individuality, coalescing into what is called a Bose Einstein condensate. The atoms are, in other words, "entangled" as one

---

*Also known as Fullerenes, after Buckminster Fuller, the American architect who designed the geodesic dome.

and, in this altogether state, can function as a superfluid or superconductor without interfering activity causing resistance to information flow.

So understanding and reproducing this process has been the overriding aim of our computing and communications industries, with the principal focus of the research on addressing ways to make this happen at warmer and, ideally, at room temperatures, thereby making it all practicable in our daily lives. Although these advances are set to revolutionize the way we live our external, material lives— resulting in communication systems of crystal clarity, quantum computers with infinitesimal memory, and, in our living rooms, real time three-dimensional images of our favorite sports stars—they carry with them even greater implications for our inner selves, and perhaps a deeper understanding of human consciousness and, indeed, reality itself.

So exactly how do these findings from the generously funded cutting edge of twenty-first-century science enlighten us about our own human reality? And why is there a reluctance to extrapolate that properties such as nonlocality or bilocation (being two places at one time) can occur in our everyday life?

Clearly, science is a long way from being able to teleport a human being. This would mean subjecting one of us to a state of complete dematerialization, before reconstituting us exactly the way we were. It would seem that we are far too complex, with far too "de-coherent" bodies, for this to be achieved in the near future. And it is questionable why we would need to do this—why not stay where we are, intact, and allow a real-time holographic image to be projected into our friend's home, while he returns the favor with his presence, and possibly his surroundings. This would undoubtedly enhance the human experience, and can be seen as important for our evolution while saving air miles and our collective carbon footprint. However, I remain unconvinced that the life of the water bear, arguably a sentient being far more successful and better adapted to this world than *Homo sapiens,* will be that enhanced by being teleported around the globe.

For me, the true message that this fine little creature passes on to us is that at the margins, at the very extremes, of our three- and four-dimensional lives, there appears a gateway to a timeless dimension of togetherness. Witness the consistency in reports from those who undergo life-changing near-death experiences. And the vivid accounts of the very young remembering past lives. And the synchronicities experienced between loved ones at times of high emotion or physical danger. These are the experiences recounted to me on a daily basis within the quiet confidentiality of my consulting rooms. Experiences that invariably seem to be linked with the compassion one person has for another. An unforced coherence and resonance between two hearts that engages, and entrains, a universal resonance that carries with it a sense of "just knowing." One could say that our complex chaotic state of entropy has been temporarily bypassed as we experience a glimpse of the symmetrical, balanced togetherness that could be genuinely described as blissful, even heavenly.

I first became intrigued by the science of entanglement between atoms precisely because it seemed to mirror the "entanglement" I was witnessing between loved ones in my medical practice. For me, this microscopic science was more the metaphor, and my shared experiences more the reality. As each of us comprise, I was reliably informed, nine billion billion billion atoms, I naturally became further intrigued to find out just how coherence could take place on such a vast scale to explain these experiences.

Until recently, the conventional view expressed by mainstream scientists has been that the experimental quantum world of the very small has precious little relevance to our day-to-day lives, and less still to the workings of our bodies. In chapter 7, we will see that this position is becoming no longer tenable, as biologists explore the warm wet world of the living, armed with the sophisticated measuring tools of twenty-first-century nanotechnology. The motives behind this pioneering research may so far still be commercial (for example, the production of solar power, perfumes, and natural therapies), but the new science of quantum biology is rapidly dispelling the myth that the quantum world can only be accessed by living beings, such as the water bear, capable of surviving temperatures approaching absolute zero.

The principal sponsors of research into teleportation, cloning, and bilocation are, understandably, the IT and communication industries. One of the largest obstacles holding up the successful development of the quantum computer and large-scale teleportation is the extremely low temperature needed to be reached to facilitate these entangled states. Ironically, it seems that the most vital clue to solving this problem may lie within the warm moist bodies of the researchers themselves. In the science fiction movie *Avatar*, these problems appear to have been solved by the middle of the twenty-second century. Humans in their state of mechanically assisted suspended animation coexist (bilocate) with their DNA—cloned avatars—in the attractive form of the blue, peace-loving, ten-foot-tall Na'vi.

Anecdotal reports of human bilocation surface from time to time in my medical practice. Three years ago, a man in his midfifties, known and loved for his kind gentle nature, became bed-bound in the terminal stages of cancer. During the days and nights of the week prior to his death, he "appeared" to several of his close friends and relatives in their homes to say good-bye and to reassure them.

Historical reports can be found in the religious and esoteric literature. On September 21, 1774, Saint Alphonsus de Liguori apparently fell into a blissful, quiet, altered state after administering mass. He was unrousable for several hours, much to the concern of his friends. On awaking, he told them that he had traveled to Rome to be at the bedside of the dying Pope Clement XIV. They subsequently discovered that the pontiff had indeed passed away, at the exact moment Alphonsus had awoken from his trance. Apparently, there were witnesses to his presence at the Pope's deathbed.[13]

In Nordic folklore, there is a phenomenon known as the Vardøgr, in which someone is witnessed to appear in a certain place in advance of their physical arrival. In a 2002 report in the speculative *Journal of Scientific Exploration*, mechanical engineer David Leiter documented two occasions in which such premonitions appeared to happen with himself as the subject.[14] One of these involved a fellow engineer and friend witnessing his appearance at work on his day off, dressed (unusually for him) in a smart suit and tie. At this precise time, he was, in fact, near his office in his car, on his way to a memorial service for his beloved aunt, attired appropriately in suit and tie. From the passenger seat, his wife had empathized with her husband about how difficult it must be on his day off to travel so close to his workplace, prompting David to visualize himself at work. The nearest his car got to his workplace that morning was a distance of five miles.

Skeptics will, of course, be at pains to point out that these stories are simply anecdotal, and apply Occam's razor, saying that such sightings are due to mistaken identity, hallucination, or simply unbridled and vivid imagination. From my own experience over the years, however, mechanical engineers, along with orthopedic surgeons and neurologists, are not the professional group most likely to be swayed by speculative flights of fancy.

So if we are to become believers in at least the principle of human bilocation, what exactly are the environment conditions that allow us to experience such quantum, nonlocal states? I am sure you have noticed certain trends.

An atmosphere of simple friendship, cordiality, peace, and relaxation appears to be a prerequisite. Vital too are the loving relationships held secure between people at those times when the thin veil of ego is lifted—in particular, when someone close is dying. This is the true environment, I suggest, that allows us as complete and complex humans, to access the zero point vacuum without, unlike the poor water bear, having to be frozen solid. In this state of emotional truth, life presents as a profound interconnected experience seamlessly linking together popes, saints, engineers, water bears—and the water lilies of Giverny. A state of harmony, balance, and compassionate intent that must be appreciated not only by the few gifted research scientists within our best universities, but also by every one of us traveling aboard the unique floating laboratory we call Earth.

Seventh Principle of the Human Hologram: Our scientific understanding of entanglement at the quantum level of the very small is leading to an understanding of "entanglement" between humans resonating through compassion with each other and with nature. Objective deductive science is complemented and enhanced by honest subjective human experience. Entanglement and the related terms nonlocality, bilocality, and multilocality are consistent with the fractal nature of our universe and the human hologram model.

Chapter 6:
The Statistics of Nonlocality—
Measuring the Immeasurable

If you want to inspire confidence, give plenty of statistics.
It does not matter that they should be accurate, or even intelligible,
as long as there is enough of them.
                                    —Lewis Carroll (1832–1898), British author

Some would say that the scientific measurement of dimensions beyond the realms of time and space, using tools developed for and existing in those limited realms, is well nigh impossible. Critics may even accuse those pioneers investigating such phenomena of living, like Lewis Carroll's Alice, in a fanciful wonderland of make-believe. Or, perhaps more ingenuously, merely tilting at windmills, a reference to the brave but misguided Don Quixote who mistook thirty or forty windmills for hulking giants ripe for slaughter. The scientists they might say are full of lofty, heroic ideas that blind them to the true realities of life.

And yet, over the past twenty-five years, there have been many carefully designed studies and meta-analyses that have suggested that, statistically at least, nonlocal phenomena such as ganzfeld telepathy,[15, 16] mental interactions with living systems,[17] psi in dreams,[18] and intercessory prayer[19, 20] occur not simply as chance findings. A prominent figure in this field of consciousness is research scientist and author Dean Radin whose landmark books *The Conscious Universe* and *Entangled Minds* are filled with balanced reports of well-designed trials critically examining psychic or psi effects. Among those complementing Radin's work is physician Larry Dossey, whose many books, such as *Healing Words,* explore the roles of intent and prayer in healing interactions.

However, the act of praying is itself something difficult both to categorize and to measure. As a youth, I well remember every morning in our chapel at Epsom College repeating parrot fashion the words of endless complex prayers, forever longing for the welcome release of the final amen (the only word of any prayer that the congregation, to a boy, voiced with any semblance of enthusiasm). I now compare that daily prescribed experience with the times later in life when I have prayed for peace in others and in myself. Times when loved ones have been suffering, dying, or simply in some kind of trouble. Or the impassioned desperate wish for comfort and relief for those in war-torn or earthquake-ravaged communities.

Despite several studies supporting the effectiveness of prayer on the healing process, a 2006 paper in the *American Heart Journal* showed no such effect on the recovery of people after heart surgery. The ten-year study involved 1,802 patients at six hospitals.[21] They were divided into three groups; a third were not prayed for, a third were told they might or might not be prayed for, and a third were told they definitely would receive a prayer. The prayers came from three congregations: Saint Paul's Monastery in Minnesota, Silent Unity in Missouri, and the Camellia Community of Teresian Carmelites in Massachusetts. All those praying were strangers to the patients. The mode of prayer was well defined: the first names of the patients and the initial of their last names were to be stated in the prayer. Although those praying were free to use their own words and methods, the following phrase was a compulsory addition: "for a successful surgery with a quick, healthy recovery and no complications." The results in the thirty days after surgery showed no difference between the prayed-for and the not-prayed-for. In addition, those who knew they were prayed for had a significantly higher (59%) incidence of complications compared with those who were uncertain (51%).

This study, of course, prompted much debate. The antagonists remained critical of any government money being spent on such futile ventures as prayer research papers. (In this case, most of the funding had come from a private source, the John Templeton Foundation.) The protagonists claimed that such a controlled enforced way of praying was both unnatural and ineffective. Dean Marak, a coauthor of the study and chaplain at the Mayo Clinic, agreed, commenting that the study said nothing about the power of personal prayer for family members and friends.[22]

The authors of the most recent (2009) Cochrane meta-analysis of the effects of intercessory prayer felt that, examined collectively, the results of the trials were at best "equivocal."[23] They concluded: "We are not convinced that further trials of this intervention should be undertaken and would prefer to see any resources available for such a trial used to investigate other questions in health care."

Compassionate heartfelt intent is difficult to measure. In chapters 14 and 16, I'll explain how the Institute of HeartMath is approaching this problem scientifically both locally for individuals and nonlocally in their Global Coherence Initiative projects. These involve many people around the world synchronizing their intent in attempts to provide peace and harmony on our planet.

My own understanding and experience of prayer for the sick leads me to support the views of Dean Marak. In most cases of the most intimate forms of person-to-person prayer, a bond of compassion has already been established between the pray-er and the pray-ee. The prayer then becomes a simple, heartfelt expression of this love; it is the pure intent behind the prayer that is the essence of the act. Or as John Bunyan, the seventeenth-century author of *Pilgrim's Progress,* observed: "In prayer it is better to have a heart without words than words without a heart."

Bunyan was describing something that can neither be measured nor controlled—something silent and intangible that exists beyond such limits. Yet it is a precious resource we discover somewhere deep within us whenever life is truly testing us. I doubt that even the most hardened skeptical soul would deny himself a prayer for his own seriously sick child based solely on the negative findings of a scientific study. However, even when the weight of statistical evidence favors nonlocal, entangled connections on a human scale, there are those who feel that the whole concept is so nonsensical that new scientific rules should apply.

In 1995, the U.S. congress asked independent scientists to evaluate the government's remote viewing program. Remote viewing is the term used to describe the process whereby an individual, often in an meditative or relaxed state, can receive information, intuitively, from a distant location on Earth. The U.S. government was interested in the potential of this method for effective espionage, especially during the Cold War. For example, in 1973, Pat Price, a gifted remote viewer, given only a set of map coordinates, identified what was later discovered to be a Soviet missile plant. He was able to describe with accuracy the shape of the machinery used.[24] One scientist, the prominent statistician Professor Jessica Utts from the University of California at Davis, discovered an accuracy rate within the remote viewing program of 34 percent, a figure that could not statistically be explained by "chance or flaws in the experiments."[25]

Even the more skeptical British psychologist Professor Richard Wiseman agreed that "by the standards of any other area of science remote viewing is proven." He concluded, however, that because remote viewing is "such an outlandish claim that will revolutionize the world, we need overwhelming evidence before we draw any conclusion." So while the debate rages on this rather outlandish subject of nonlocality within the lofty, but often chilly, ivory towers of academia, I continue to be educated every week by the stories I hear in the warm comfort of my consulting rooms. Fascinating tales of synchronicity—commonly shared nonlocal experiences that effectively secure lasting bonds between loved ones.

While I was writing this chapter, Julie, a woman in her late forties, came to me with longstanding pain in her foot and back. She managed a thriving business with her husband while meeting the ever-present demands of her three teenage children. She explained that before she met her husband, she gave birth to a son, now aged thirty, who was adopted out. For many years, protective of her other children and husband, she was reluctant to make contact with her son. Five years ago, however, prompted and supported by her husband, she decided to contact the adoption agency in an attempt to find her son.

She duly made an appointment and underwent a searching interview with one of their senior staff members. She was warned it was most likely that her twenty-five-year-old son would prefer not to make contact at this time, as he had not instigated such a meeting. She was reassured, though, that it was likely that some time in the future he would feel comfortable about reuniting with his birth mother, especially if he knew that she was keen. Just as the report was being completed on the afternoon of the interview, the agency received a call from a young man named Bradley requesting a meeting with his birth mother, who, it was quickly discovered, was Julie. This was the very first time he had made contact with the agency. All parties have subsequently confirmed that this was truly a synchronous event. Not only have Julie and Bradley reunited in a truly life-changing way, but also Bradley's adopted family have become close friends with Julie's family.

Statistics will never prove conclusively whether this joyful reunion was a synchronous, nonlocal event facilitated by shared focused consciousness within an environment of truly unconditional love, or merely a happy coincidence. And really, does it matter? After all, for Bradley, Julie, and their respective families, it is the joy of the experience, and the effect on their future lives, that is significant—far beyond words, statistics, and academic debate.

However, mainstream academics' acceptance of the science of such nonlocal phenomena will empower many to follow their intuition, seek compassion, and find creative solutions to their problems. And, as we have already discovered, there is one branch of science, evolving rapidly in these early years of the twenty-first century, that may well promote such an acceptance and help bridge the gap between the opposing world views expressed here. It carries the title of *quantum biology,* and there are growing numbers of scientists who speculate that its implications could, indeed, to use Professor Wiseman's phrase, "revolutionize the world."

Chapter 7:
Quantum Biology—
The Science that Came Out of the Cold

*Every great advance in natural knowledge has involved the
absolute rejection of authority.*
> —Thomas Huxley (1825–1895), British biologist

Necessity, who is the mother of invention.
> —Plato (c. 428–348 BC), Greek philosopher

One of the greatest challenges facing us in the twenty-first century is how
we are to adapt as a species to the rapid changes in our world's climate. It is
clear that greenhouse gases are accumulating in our atmosphere, while our
earth's lungs, its forestation, are being destroyed to meet the consumer needs
of a rapidly expanding human population. Our trees and greenery are vital to
our survival; they absorb harmful excesses of carbon dioxide and, through an
interaction with sunlight from the sky and water from the ground, produce
sugars to help them grow and oxygen for us all. This life-giving process is
known as *photosynthesis.*

Of course, we must approach the problem of climate change from many
angles. When I was born in 1951, the world's population stood at a little over
2.5 billion. In 2010, it reached seven billion, with projections predicting yet
another billion humans inhabiting our planet by 2020.* So managing this
growth is becoming a priority, a practical and philosophical issue of mammoth
proportions as this growth largely results from human beings living longer.
It is clear, however, that we are compounding the problem by destroying our
rain forests. We are not simply biting off the hand that feeds us; it is a hand
that also, most generously and effectively, cleanses us by disposing of our
poisonous waste.

---

* 2004 UN records and projections of world population from 1800 to 2100.

One way scientists can be proactive in combating the causes of climate change is by studying just how and why photosynthesis is so much more efficient than any man-made solar energy devices, which lose as much of 20 percent of valuable light energy in the process. Research into the mysteries of plant photosynthesis could bear fruit in the form of devices that make considerably cleaner, cheaper, and more sustainable use of our sun's energy.

A dramatic breakthrough came in 2007 with the work by scientists at the University of California at Berkeley and at the University of Washington in St. Louis. The team, led by biophysicists Gregory Engel and Graham Fleming, studied photosynthesis in the proteins of green sulphur bacteria, using light from the highly specialized ultra-fast femtosecond laser. Pulses are emitted from this laser at a staggering speed—every femtosecond or 10-15of a second—just enough time to set single electrons spinning. To put that into perspective, a femtosecond is to a second what a second is to about 31.7 million years![26]

Using modern nanotechnology, they discovered something remarkable about how light is transmitted and processed deep within this bacterium. The light from the laser was transmitted by the antenna-like protein scaffolding from the surface to reaction centers deep within the cells. But rather than following one direct route as the arrows on our high school photosynthesis diagrams show, the light appeared to travel in several directions *simultaneously*. It literally appeared in several receptor sites at the same time, *entangled* in a way previously only demonstrated in the dry solid-state conditions in the quantum physics laboratories. The subsequent research paper published in *Nature* is already being hailed a landmark study, demonstrating that quantum processes are truly at work in the wet conditions of a living system.[27] The researchers theorize that the solar energy is engaged in an instant quantum "scanning" process, the purpose of which is to locate the best complete pathway for energy conversion. Only then, *after the event* and once the most efficient pathway has been "selected," does the quantum process collapse into a formal linear pattern. Some have labeled this natural selection of the "fittest" pathway *quantum Darwinism*.

However, these observations were still made at the ultra-low temperatures of previous studies. To allow this solar energy technology to become truly practical, it was important to show that this nonlocal effect (now becoming known as *photosynthetic coherence*) can happen, and be reproduced, at everyday temperatures. In February 2010, University of Toronto biophysicist Greg Scholes published research in *Nature* that demonstrated precisely the same quantum processing of light, this time in marine algae, but most important, at room temperature.[28] At the time of writing, further research from Engel's group is confirming this in other photosynthesizing organisms, strongly suggesting that the quantum world exists at the most fundamental level of life itself. So with this research—an exciting twenty-first-century synthesis of nanotechnology, ecology, and biology—aimed at exploring ways we can best save our planet, we are gaining remarkable insights into the true science of life.

Research into quantum biology is being funded not only by the solar energy industry. IT companies are obviously keen to explore the possibility of quantum computation at the temperatures in which we live. They, like others, however, are beginning to realize that quantum processes within a living being represent a holistic mix that may not prove easy to replicate with technology. As demonstrated by Engel and Fleming's work on bacteria and Scholes's research on common marine algae, there exists a state of harmony and balance between the organism, its environment, and, as discussed in chapter 3, our own observation. In fact, these three elements are probably inseparable—intimately and primarily connected in a state of entanglement.

But is there hard scientific evidence that such quantum activity exists within our own bodies? Where should we look and, more to the point, who is motivated enough and has the financial means to support such sophisticated and expensive research?

It could be said that the answers to these questions are turning out to be right inside, if not under, our noses. Our sense of smell has forever provoked powerful emotions and, for centuries, has been known to communicate fundamental truths in a way even more effective than the spoken word. Shakespeare's Juliet seemed to share this conventional wisdom as she stood on her balcony pondering just why it could be that a young lad named Romeo affected her so, while softly uttering the immortal lines: "What's in a name? That which we call a rose by any other word would smell as sweet."

Smells can transcend the smeller beyond the constraints of time and space, and help awaken deep intuitive feelings. Helen Keller, the deaf-blind twentieth-century social activist, described smell as "a potent wizard that transports you across thousands of miles and all the years you have lived." The great screen actor Paul Newman once explained how he selected the movies in which he was to star. It was not simply the words of a script that would convince him he was exactly right for a role. It was as important for him to "see colors, imagery. It has to have a smell. It's like falling in love. You can't give a reason why."

But now, in the early twenty-first century, there are scientists who are very keen to find a reason why. And helping fund this work is the Pentagon's research outfit DARPA (Defense Advanced Research Projects Agency), keen to develop sensing devices that can identify security threats through solid objects.[29] By engaging a team of researchers at the Massachusetts Institute of Technology, in a project named MITRealNose, they hope to develop a sophisticated e-nose as sensitive as that of a sniffer dog. And helping this enterprising team is biophysicist Luca Turin, founder of the fragrance company Flexitral.[30]

In 1996, it was Turin who challenged mainstream thought by theorizing that the chemicals of smell, known as pheromones, could not produce their evocative effects simply by chemically locking onto receptor sites in our noses.[31] If this lock- and-key model was correct, then very similarly shaped molecules would have to produce similar smells. And in Turin's experience, this was far from the case. Instead, he proposed that quantum processes were at work, with instant scanning and connections entangling the smell with the smeller. He proposed that vibrations from the interaction of the smell and the nose receptors instantly set into motion a unique pattern of vibration throughout all the atoms of the odorant.

In 2007, Turin's theory gained strong scientific support from four physicists at University College London.[32] The group confirmed that Turin's model follows the laws of physics, if one assumes that there is at play a process known as *quantum tunneling*. In simple terms, this is a state of entanglement, well known and demonstrated in nanotechnology, that allows information to "tunnel" through physical objects instantly. The researchers deduced that it was the unique vibrational pattern of the odor molecule, not its static shape, that on meeting a receptor, set in motion this quantum tunneling, with odor molecules and receptors resonating with each other instantly as if engaged in an exotic, frenzied dance. It was this overall pattern of vibration that could then be relayed to the brain via the sensory nerve of the nose, the olfactory nerve.

This research into the quantum foundations of life is at its infancy. It has been spawned by two of the great challenges we face in the early twenty-first century: climate change, and national security in response to terrorism and drug trafficking. It has only been made possible because of the rapid advances in information technology, and the relentless drive by scientists towards that holy grail of IT: the quantum computer. Perhaps not surprisingly, this research has also revealed how utterly essential the quantum world is to the way plants and animals interact with, and connect to, their environment. Further studies are showing that quantum "compasses" may be operating in the eyes of migratory birds, allowing them to sense, even see, magnetic fields that guide them on their vast journeys around the globe.[33, 34]

Biophysicists are also beginning to discover that nonlocal entangled states must be at play throughout our bodies. Approaching our internal workings with a deep knowledge of mathematics and physics, they have deduced that without this vital quantum realm, many of the body's chemical reactions would be unsustainable, losing too much energy in the process. For instance, calculations and measurements now show that quantum biology must be behind many chemical reactions happening within us every moment, in particular:

1.   How green tea, an oxidant, neutralizes free radicals instantly and synchronously in our bodies[35]

2.   How DNA, water, and protein molecules communicate as genetic information is translated into the making of the body's building blocks, proteins[36, 37]

It is also clear that these living quantum properties are only possible because of the subtle, yet essential state of balance held by the organism in resonance with its natural environment.

It is likely that, as this century progresses, scientific advances will lead to the manufacture of super-fast quantum computers with near-unlimited storage and memory. However, the coherence so vital to the life of an organism—plant or animal, large or small—cannot be created simply by assembling all its separate component parts, as the fictional Dr. Frankenstein discovered, to his considerable chagrin.

It was Thomas Huxley, quoted at the beginning of this section and known to the Victorian public as Charles Darwin's fearless advocate or "bulldog," who helped popularize Darwin's radical new theory of evolution. He predicted a time when we would hold in our hands the scientific tools with which we could create human life, and thereby control our destiny. That time is now with us; we have entered an era that carries with it an awesome responsibility for the future. Ethics and morals of the highest standard are essential. It was Huxley's very own grandson, Aldous, who, through the classic novel *Brave New World,* had the foresight to caution the twentieth-century public about the potential dangers of eugenics, the genetic engineering of a master race.

Quantum biology is still in its infancy. As it grows and matures, it will play an important role in convincing the academic world that we are indeed multidimensional beings, living in harmony with a multidimensional universe. It will provide comfort and confidence to those who already know in their hearts that this is so. By opening minds and hearts, we can discover new ways to heal ourselves and our planet. For we can't heal one without healing the other.

Eighth Principle of the Human Hologram: In the early twenty-first century, scientists are just only beginning to explore the quantum foundations of life.

Chapter 8:
Of Ants and Men—Life in the Colonies

Turn on the prudent ant thy heedful eyes. Observe her labors, sluggard, and be wise.
—Samuel Johnson (1709–1784), British author

Ant-watching is one of my favorite pastimes. When in a reflective mood, there is little I like more than turning my attention downwards to the world that exists in miniature beneath our feet. Unlike bird-watching, finding subjects requires very little effort; in fact, they often come to us, especially if the sugar jar is left open. Ant-watching can be enjoyed equally out of doors or in the warm comfort of the home, while lounging in a deckchair, armchair, or even on the toilet. And no need for expensive binoculars—only, for the more mature eyes like my own, bifocals.

The first thing the novice ant-watcher will notice is that, not only are these tiny creatures tirelessly diligent, they also appear, to their fellow ants at least, very polite. When they are traveling on their highways, there is no sign of road rage. In fact, a procession of ants moving in a straight line in the same direction slows down and speeds up in a precisely synchronized fashion. There is no bunching up or colliding—no traffic jams. If more ants join a platoon, the line simply gets longer.

This model behavior was recorded by scientists in a joint 2009 study from institutions in Germany, India, and Japan.[38] It has naturally captured the attention of those whose job it is to help control traffic in our cities. Whilst it may take some time before humans evolve away from their egocentric behavior on the road, one of the researchers, Andreas Schladschneider, from the University of Bonn in Germany, predicts that in "the future, our cars might be connected electronically and transmit information about velocity changes immediately."

It is thought that ants do this chemically by forming trails with their pheromones. Desert ants, however, are known to navigate their way up to one hundred meters away from their home without visual landmarks, and away from pheromone trails. Using twin antennas, they respond to "odor signatures" from the habitat, literally smelling the scenery in stereo.[39] (We too are thought to smell in stereo—ever wondered why, in addition to owning two eyes and two ears, we have two nostrils?)

There have been no studies or calculations to date to confirm that quantum processes are at work linking ants to their environment and to each other. It seems logical, however, to suggest that Luca Turin's theory of smell, as supported by the physicists at University College London (see chapter 7), could apply equally here. That is, at the roots of this extremely efficient and sophisticated navigation behavior and at the roots of the amazing synchrony of movement within the ant colony, lie vibrationary and quantum activity. There is one remarkable ability, common to ants, bees, and bacteria, that continues to intrigue biologists, physicists, and pharmacologists. *Quorum sensing* is the term given to a skill inherent in an individual organism that allows it to sense the size of its colony, and then respond, appropriately and instinctively, to this information.

For example, whenever a nest is destroyed, ants scatter hither and thither, searching out various potential sites, to return at intervals to the destroyed nest. Once there, they are keen to return quickly with fellow workers to the best new site and, as a result, the numbers at this new site increase more quickly than others. Once a critical mass (or quorum) of ants is formed at a new site, it is sensed by the ants who travel back and transport their young, their queen, and fellow workers to the new nest. They even recruit the remaining ants still looking for other sites. So even though each ant hasn't checked out all potential options, a group consensus is reached through democratic means. Through this process, the ants eventually congregate in their new home.

Honeybees too have the ability to quorum sense, communicating to their fellow worker bees about the excellence of a potential new site through an exotic maneuver known as a waggle dance. The longer the bee dances, the more promising it feels is the new location. This dance is also used to indicate just where the best pollen and nectar is to be found—with the length of the performance and the angle of the dance to the sun indicating distance and direction, respectively.

But it is the quorum-sensing skills of bacteria that are generating the most study. Resistance to antibiotics, due mainly to overuse of the drugs, has become a serious health problem over the past twenty years, so new ways are being sought to treat and prevent bacterial illness. To quote the January 2009 *New England Journal of Medicine:* "We have come almost full circle and arrived at a point as frightening as the preantibiotic era: for patients infected with multidrug-resistant bacteria, there is no magic bullet."[40]

One solution to this health crisis may result from a deeper understanding of how bacteria behave as entangled networks. Once bacteria reach a certain population, they start behaving collectively as a colony, or biofield. This is seen most dramatically in a process known as bioluminescence, a spontaneous emission of light by all the bacteria in the colony. Benefiting greatly from this is the Hawaiian bobtail squid, in whose body reside millions of such bacteria (*Vibrio fischeri*). The squid tends to swim near the surface of the ocean, laying it open to attack from predators below. At certain times of the day and night, however, the bacteria spontaneously light up, precisely matching the light from the sun and moon, effectively camouflaging their host.

The dramatic display of bioluminescence from another marine bacterium, *Vibrio harveyi,* has even been photographed from outer space.* In 2005, photographs taken from an orbiting satellite clearly show a cigar-shaped patch stretching over one hundred miles in the northwestern waters of the Indian Ocean. This phenomenon, known traditionally to seafarers as a "milky sea" (the light is actually blue rather than white), was vividly described by Jules Verne in his science fiction novel *20,000 Leagues Under the Sea.* Captain Nemo's submarine Nautilus encountered a mysterious "sea of milk" as it traveled, half-submerged, on its perilous journey through the Bay of Bengal.[41a]

---

*Yet another bioluminescent marine organism, the sea pansy, is helping medical researchers achieve better imaging of our internal organs. By injecting the enzyme responsible for illuminating the sea pansy together with tiny nano-sized synthetic crystals known as quantum dots, crystal-clear images of previously hidden parts of the body light up and can be recorded easily. The overriding limitation of injecting quantum dots alone has been that an external light source is required. Not so with this self-generating light-emitting hybrid team of nature and cutting-edge technology.[41b] Interestingly, quantum dots exposed to laser beams have now been shown to entangle photons of light, leading one to speculate whether bioluminescent organisms have their own natural and intrinsic version of these tiny crystals.

Returning now to our twenty-first century, it is the resistance of antibiotics to harmful bacteria, and the elusive cure for cancer, that is forcing medical researchers to delve beyond the constraints of their restrictive, linear models. Leading scientists are calling for a broader, less reductionist approach if we are to begin to find lasting solutions. Renowned microbiologist Professor Julian Davies from the University of British Columbia encourages more research into how bacteria use signals as they communicate within their "mosaics of interactive networks."[42] The cosmologist, biologist, and physicist Professor Paul Davies of the University of Arizona is pioneering research into the physical properties of cancer cells. He explains: "We look at the forces that act upon them, look at their mechanical properties, their electrical properties, how they cluster, how they act as communities."[43]

It would appear that scientists are taking their first tentative steps towards a deeper, more holistic, understanding of how a massive colony of cells functions in perfect harmony within a living system such as the human body. By studying the outer world of bacterial and insect biofields, valuable insights can be gained about the inner world of our bodies. So far, though, the focus has been on cell-to-cell communication via local chemicals and electrical forces. There is still an overriding reluctance to see past these intermediate steps and to ask the seemingly obvious question: What is the fundamental system that coordinates trillions of individual cells, whether within an ant colony or a human body, allowing them to grow and function in such harmony?

Using David Bohm's philosophical concepts, we are still focusing on the explicate order of surface local activity rather than the less visible implicate organization that lies beyond the reach of our senses. It is, to me, akin to describing a fascinating game of chess merely in terms of the mechanical movement of pieces over a board.

In his landmark book of 1925, The Soul of the White Ant, South African biologist and poet Eugene Marais detailed, in his native Afrikaans, his meticulous research into the community life of the termite. A highly original systems thinker, he compared the whole termite colony and nest with the body of an animal. To Marais, the workers and soldiers represented red and white cells, the fungus gardens the digestive organ, and the queen the brain "controlling the collective mind." He described the sexual flight of the kings and queens as being identical to the escape of spermatozoa and ova.[44]

In one experiment, he divided a termite mound into two with a thick metal plate, observing the activity of termites on either side of the barrier. The plate effectively blocked electrical and pheromone transmission (it is worth noting that termites are completely blind). Despite this, the two sides of the mound grew in perfect symmetry, suggesting connections between the ants via a field beyond electricity and chemistry. He also found that if the queen was removed from her cell and destroyed, the termite activity completely ceased. Prior to this, even if she was separated from the workers by the steel plate, their meticulous work would continue.

Although this experiment has not been repeated in modern times, if confirmed, it would lend credence to evolutionary biologist Rupert Sheldrake's theory of *morphic resonance,* whereby the behavior and growth of organisms is organized via a fundamental field of information. And to Messrs. Bekenstein, Susskind, and 't Hooft's theory that our physical world is merely a holographic projection, a virtual reality created by our senses from a mysterious invisible realm.

Unfortunately, Eugene Marais did not live to see his theories accepted, nor his contribution to science fully recognized. His work on termites was plagiarized by Nobel Laureate Maurice Maeterlinck, who expounded Marais' theories as his own. And in 1936, Marais, a long-time morphine addict prone to bouts of deep melancholy, shot himself in the head, after attempting unsuccessfully to end his life a few moments earlier by firing a shotgun into his chest.

Chapter 9:
The Patterns of Nature

It is true that a mathematician who is not somewhat of a poet will never be a perfect mathematician.
—Karl Weierstrass (1815–1897), German mathematician, discoverer of the first fractal

Big whirls have little whirls, That feed on their velocity; And little whirls have lesser whirls, And so on to viscosity.
—Lewis Fry Richardson (1881–1953), British mathematician

The computer age has not only flooded us all with information, it has also created an exciting new medium for artists. This eclectic mix of mathematics, technology, and image has allowed the science and appreciation of fractal geometry to flourish. However, it took the raw genius of Benoit Mandelbrot and the vision of his U.S. employer, the computer giant IBM, to introduce the modern world to the beauty and complexity of its form.

Mandelbrot was born into a Lithuanian Jewish family in Warsaw in 1924. Under threat of a Nazi invasion, the family moved to Paris and then, after its fall in 1940, to the south of France. It was here that he became fascinated by the shapes of nature. He found it intriguing that estuaries of rivers mimicked the shape of human blood vessels; and that a small piece of cauliflower mimicked the whole vegetable. The disruptions to his home life meant he received only limited formal schooling; apparently, he never learned the alphabet or multiplication past his five times table. So visual was his memory, however, that he was able to pass mathematic exams simply by visualizing the shape of the problems.

Among the many visionary scientists that inspired Mandelbrot was the little-known Lewis Fry Richardson, a British pacifist and naturalist who had studied weather patterns, in particular the "whirl within whirl" turbulence of wind. He was similarly fascinated by other unsolved mysteries of the world around him. For instance, he was the first to pose the question: "Just how long is the coastline of Britain?"

If we motor around a country, taking all the coastal routes, the odometer will clearly tell us the mileage we have covered. This only reflects the distance traveled by road, however, and is in no way a true measurement of the coastline with all its intricate nooks and crannies. In fact, if we were to examine the surface of every rock, and every sea wall, at the point that they met the sea's high-water mark, we would discover countless tiny crevices within crevices. If we were to use a modern microscope, even more intricate patterns would be viewed. In fact, the circumference of a land will increase proportionally depending on just how minutely we examine it; on the subatomic level, it would approach infinity!

Richardson's observations proved a guiding inspiration to Benoit Mandelbrot. During his years at IBM, Mandelbrot studied the mathematics and geometry of the natural world, applying computer imagery to the theories of other nineteenth- and twentieth-century mathematicians. The result was the now-famous Mandelbrot set, the periphery of which reflects a complex but self-replicating fractal pattern. As one zooms into the border of a set with more and more powerful computer microscopy, beautiful new worlds appear: the self-repeating Julia sets reminiscent of fire-breathing dragons. Mathematically, there is still much to learn about the Mandelbrot set, but encoded in it is both the symmetry and the chaos of the natural world.

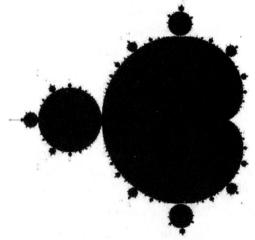

Figure 10. Mandelbrot set.

We can begin to understand this balance as we observe how the fractal pattern of each individual snowflake is unique, subtly different from all others. Identical twins, however alike they are, always have different fingerprints. It is as if, although an underlying symmetry lies at the very roots of nature, it is the individual expression of chaos and diversity in the physical dimension that is an equally vital component of life. As if by exploring the very margins of our existence—by studying a coastline, a Mandelbrot set, or even a cauliflower—we discover just why we are all so diverse, and why we are exposed to an infinite number of choices in our lives.

Our bodies are fractal; like the limbs of a tree, our blood vessels branch out from a central trunk, our aorta, to smaller and smaller vessels that infiltrate our tissue, providing nutrients and removing waste from every part. Our lungs comprise ever-smaller fractal branches allowing greater and more efficient absorption of oxygen into our blood. Our brains too are fractal, with all their bizarre folds, wrinkles, and complex intercommunicating networks of one hundred billion nerve cells.

Fractal science also explains just how over six feet of DNA can be packed into each of our cells to form our human genome. Although we are accustomed to seeing pairs of chromosomes about to divide in neat X-shaped forms, at all other times they are ball shaped, with the DNA packed in, to quote computer scientist Erez Lieberman-Aiden, "like an incredibly dense noodle ball."[45] This is not simply the most efficient, knot-free way to package DNA, it is perfect too for the cross-communication needed between different parts of the molecule. Its fractal packaging is also likely to be highly important for DNA's nonlocal connections to quantum realms.

But perhaps the most intriguing, and hitherto neglected, fractal system within our bodies is our cytoskeleton. This network of connective tissue not only infiltrates every cell of our body, but also makes up a staggering 70 percent of each cell's mass. It provides the internal scaffolding for our tissue, the spindles that pull our chromosomes apart, and the semiconductive electrical circuitry known popularly as our Chinese meridian system.[46]

Recent evidence shows this fibrous, fractal network plays a major role in "the spatial organization and regulation of translation, at both the global and local level, in a manner that is crucial for cellular growth, proliferation, and function."[47] In other words, our cytoskeleton acts as a living map upon which we grow. But it is likely there is even more to the cytoskeleton than this. According to Sir Roger Penrose and Professor Stuart Hameroff, the structures within the cells known as microtubules— some antenna-like, others with a triple-helical shape—may act as microscopic foundries within whose walls human consciousness is processed. Just as a tree employs quantum processing within its fresh green leaves converting light to energy, at the distant tips of the most distant branches of our own cytoskeleton, quantum biology may also be at work. According to the Penrose-Hameroff theory, it is through these structures that the quantum realms are processed, where the probability waves collapse down into the reality, or maybe the illusion, we call life.

Microtubules are simply everywhere in the human body, and in the bodies of all animals evolved over the past 540 million years. They are particularly numerous in the cells of our brain; and it is likely, according to Hameroff, an anaesthetist in Tucson, Arizona, that this is where general anaesthetic agents do their work, altering our consciousness to make us unaware of any pain and oblivious to the passage of time during a surgical operation.

Fractal geometry, the study of the patterns of nature, is taking us closer to understanding the roots of our existence. It represents a further shift away from a science that tries to describe our universe in strictly linear, mechanical terms. The mathematicians and scientists who pioneer this work, like gifted poets, observe and translate the rhythmic language of life and, like great artists, unveil its hidden beauty. They personify an evolution of human consciousness by seamlessly blending the analytical with the conceptual, applying, metaphorically at least, both left and right sides of their brains. Furthermore, they allow modern cutting-edge technology to aid, confirm, and advance the wise observations and theories of their predecessors.

It could be said that the true nature of our world is revealing itself to those who are ready.

Ninth Principle of the Human Hologram: The fractal nature of our universe, consistent with holographic theory, is now a recognized tenet of mainstream science.

Chapter 10:
The Human Biohologram—A Working Model

Natural science does not simply describe and explain nature; it is part of the interplay between nature and ourselves.
—Werner Heisenberg (1901–1976), German physicist

We are enfolded in the universe.
—David Bohm (1917–1992), American-born British physicist

The scientific world has yet to entertain fully the possibility that we ourselves are cocreators, and reflectors, of our world around us. The demand for objectivism at all costs still overrules the voice of pure science that tells us that the subject and object, the observer and the observed, are never completely separate. Early in the twenty-first century, few scientists seem to have grasped the significance of the words of the late Danish physicist Niels Bohr: "A physicist is just an atom's way of looking at itself."

Looking at, and within, ourselves is never easy. As a doctor, I know I have always proved to be my own most difficult patient, forever struggling with the time-honored advice "Physician, heal thyself." But if we are to accept that our universe is holographic, that fields of information or consciousness lie fundamentally behind all we perceive, we must also come to terms with the fact that all this applies to our own being. This could leave us feeling fragile, vulnerable, and extremely unstable—not merely in unfamiliar territory, but precariously lost in space.

But I am convinced that if we do embrace ourselves as human holograms, our lives will be enriched beyond measure. Soon (in section three) I will explain how understanding this dimension of ourselves can take us to new levels of personal health, without the need for excessive and expensive medications or equipment. This is the motivation behind my writing this book.

Even though our understanding of holographic science, fractals, and quantum biology is still in its early infancy, there are a few respected scientists leading the way with groundbreaking research. Their work cited here is still regarded as highly controversial, but there have been no successful attempts to discredit their findings. This is partly because few scientists have the expertise to challenge the extremely specialized nature of their research. But also, quite possibly, the living quantum effects observed and measured by a believing scientist with positive intent might differ from those of a skeptical investigator.

In the early 1970s, Fritz-Albert Popp first detected a special "light" emitted and transmitted by living cells. He named this subtle energy *biophotonic* energy and the particles that were involved *biophotons*.[48] Through sensitive electronic devices, he succeeded in measuring this energy as it was emitted from both plant and animal tissue. Other researchers have studied its wave-like properties in an attempt to understand just where this energy fits within the whole spectrum of electromagnetic energy. Interestingly, the consensus is that it most closely resembles the widely spaced waves of sound, rather than the narrower waves of light. Some have speculated that this allows the "beams" of biophotons to travel through the body more effectively with less resistance.

Popp's experiments have revealed fascinating insights into the nature of this subtle, yet vital energy. He and other physicists discovered that cancer cells emit more biophotonic energy than healthy tissue, suggesting that this is truly a sustainable and renewable resource in healthy tissue. Healthy cells are less inclined to lose energy, actively recycling it. It appears that the rampant, uncontrolled multiplication of cancer cells outpaces the body's ability to reabsorb this energy,[49–53] leaving the sufferer of cancer drained and weak. Another study revealed that someone in a balanced, calm, meditative state emits even less of this energy. In other words, in this state we are more likely to conserve our vital energy, with resulting health benefits.

There is also speculation that biophotonic energy is processed by the DNA within the nucleus of each cell,[54] and then fired out of the cell in a coherent laser beam with "the trajectory of a bullet." Others have speculated that the body's connective tissue, in particular the helical-shaped microtubules, flagellae, and collagen molecules, is involved in propagating these beams of biophotons within the body.

Just as in a laser light show, our own internal beams of biophotons interact and interfere with each other, forming a truly holographic field or matrix. Only in our body, this show is complex beyond our wildest imaginings, with many trillion sources all firing at once. It is even envisaged that our growth and regeneration unfolds within this hidden matrix, which when we are in perfect health, is in perfect harmony with our natural environment.

Detailed theoretical models of this process have been proposed over the past twenty years by pioneering, broad-minded scientists. Physicist Bevan Reid describes a step-by-step process whereby the "virtual" energy of pure space converts initially into these beams of biophotons, which, in turn, form the foundational matrix for living matter.[55]

But before we describe this fascinating and complex act of conversion occurring at every given moment within all of us, maybe we should examine the special, sharing relationship we all have with the space between us. Scientists are now beginning to understand just how and why our physical body of cells, our hardware, is in constant communication with this mysterious, ubiquitous, "virtual" energy of space.

But surely space, we have been led to believe, consists of simply nothing, the empty area into which everything else fits? Not so, according to quantum physicist and one-time colleague of Albert Einstein, John Wheeler. All space, including the proportionally vast areas within our atoms where no particles exist, comprises an infinite number of tiny spiral vortices, or wormholes, collectively known as the "quantum foam."[56] Each one, according to this theory, acts as a portal to the entire library of universal information, each one a mini star-gate to other dimensions.

Reid proposed that this space energy "stored" within the vacuum of a symmetrical spiral makes contact with our dense living tissue, and then converts into the coherent laser-like beams I have already described. These beams of biophotons then contain both the infinite store of information from space, and also further information acquired from this collision with the physical hardware of our bodies. Our physical bodies are wonderfully efficient memory storages devices, complex computers containing the history of everything we have experienced. So it is proposed that space energy, on making contact with our bodies—our uniquely personal database of experiences—instantly downloads this vital information, storing it in its own vast files for posterity.

There is therefore a constant and harmonious exchange of information between each of us, and indeed all sentient beings, and the invisible field of universal consciousness. This process is known as *resonance*.[57] So space energy is being continuously updated and upgraded by contact with our bodies, meticulously recording in its vaults all our earthly experiences. This record is enriched beyond measure by that special gift so valued by all human beings: the *free will* with which we are all so generously entrusted.

So every deliberate act, every thought even, in every one of our unique lives contributes in a significant way to this universal, timeless *field of consciousness*. As each of us evolves spiritually, so in turn does the *field*. If we fully embrace this model, we human beings inherit an awesome level of responsibility. We are clearly, it would appear, cocreators of our own, our planet's, and our universe's destiny (see figure 11).

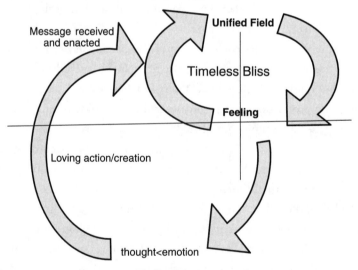

Figure 11. Challenges in time and space.

It is our own bodies that also stand to benefit immeasurably in this cooperative venture. To explain this, let's now revisit the intricate process whereby interfering beams of biophotonic energy (another name for *biophotons is bosons*), derived from space energy and processed through our DNA and microtubules, form a field or matrix within which our physical tissue manifests (see figure 12).

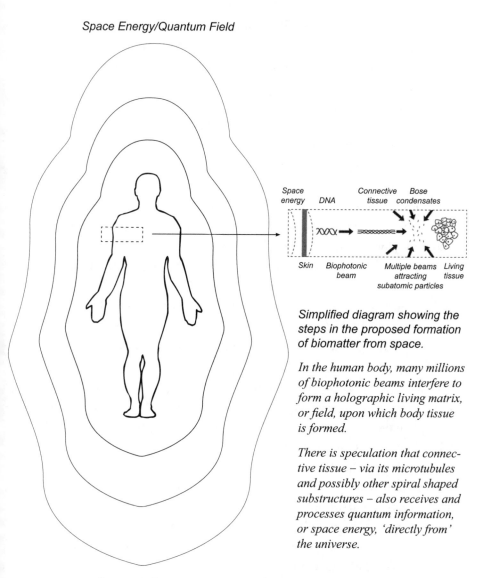

Space Energy/Quantum Field

Space   Connective   Bose
energy   DNA   tissue   condensates

Skin   Biophotonic   Multiple beams   Living
           beam   attracting   tissue
                     subatomic particles

Simplified diagram showing the
steps in the proposed formation
of biomatter from space.

*In the human body, many millions
of biophotonic beams interfere to
form a holographic living matrix,
or field, upon which body tissue
is formed.*

*There is speculation that connective tissue – via its microtubules
and possibly other spiral shaped
substructures – also receives and
processes quantum information,
or space energy, 'directly from'
the universe.*

Figure 12. Bosons interacting with microtubules and cells.

Scientists have recently discovered that electrons and other subatomic particles are attracted to this field, condensing onto and within its matrix, in precisely the way a cold glass attracts water onto its surface. The resulting mixture of particles is known as a *Bose condensate*. Bose condensates only occur in the laboratory at very cold temperatures, near absolute zero.

It is speculated, however, that a similar process may occur in the perfectly balanced, warm, wet environment of living organisms, forming the very living tissue—our structural proteins—that we can see and feel with our senses. (As we learned in chapter 7, we are now finding evidence that quantum processes occur within living organisms.)

In a bold set of experiments between 2001 and 2005, Russian scientist Peter Gariaev and his team at the Quantum Genetics Institute in Moscow demonstrated many of these effects in the laboratory. In one experiment, the pancreases of poisoned rats were completely remodeled into healthy tissue using a laser beam from a "quantum biocomputer." This machine had previously scanned the pancreas of a healthy rat of the same species, recording quantum information emitted by this healthy tissue's DNA. The information in the form of a laser beam was then directed back into the poisoned pancreases of the diseased rats. Ninety percent of these fully recovered, with completely healed pancreases, compared to a control group in which all died.[58]

Earlier in 1985, Gariaev demonstrated the so-called phantom DNA effect. After firing coherent beams of photons at DNA in a vacuum, a ghost-like hologram of the DNA appeared, which remained for some time, even after the DNA molecule was removed. He deduced that a DNA molecule was capable of receiving, processing, and projecting light. His work has since focused on the wave-like and quantum properties of the genome, exposing the limitations of the purely linear mechanical approach to unraveling our genetic code. With the help of linguistic experts, his team discovered that a language was encoded within the sequence of base pairs in the DNA that followed all the rules of our world's languages. So rather than the base pairs simply existing as complex, meaningless lines of letters, their subtle relationship to each other also carried information. In a way, sentences, even the rhythm of poetry, could be encoded in these sequences.

What's more, according to Gariaev, the 98 percent of DNA that had been dismissed as "junk" by those unraveling the genome in a linear fashion likely played a vital role in this instant translation and broadcast of information. Nonlinear, nonlocal processes were proposed to be at work both within each DNA molecule and between DNA throughout the body. An understanding of the fractal nature of life, the awareness of ever-present self-similar patterns of varying degrees of complexity and of the properties of a hologram now familiar to us, allows us to take the next step: that the information processed within these microscopic structures can be projected instantly to the whole organism.

It is this that is perhaps the most important tenet of the new science of the human hologram. As John Wheeler claimed: "No one will be considered scientifically literate tomorrow, who is not familiar with fractals." One of the most exciting aspects of this new degree of scientific literacy is that it opens the field up to disciplines that have hitherto lain well outside the formal scientific model. An appreciation of patterns has largely been associated with those with a healthily developed right brain. So, as I proposed in the previous chapter, we are witnessing a balancing of the deductive and linear, a function primarily of our left brain, with our more artistic and creative right brain.

As an example of this collaboration, in 2009, researchers from the Queensland University of Technology and the University of South Florida joined together in a study to better understand how we combine words to create meaning.[59] It appears that, as a means of expressing ourselves and conveying information to others, we pluck "out of thin air" words that are intimately associated, or entangled, with other words. In this way, we are able to convey complex ideas using relatively few words. This emerging field is becoming known formally as *quantum cognition.*

There is, perhaps, still a leap of faith needed for us to entertain such a radical new vision of ourselves as human bioholograms—existing as holographic projections constantly forming on and interacting with space. The inevitable conclusion, if this is indeed accepted, is that we live, mechanically at least, as virtual beings within a virtual reality. That the chaos and entropy we encounter in our physical lives here on Earth are simply the means by which we learn to grow consciously, both as individuals and as a collective. And that as we grow, so too does the field of consciousness—as, indeed, we are one and the same.

Chapter 11:
Science, Faith, and Evolution
in the Twenty-First Century

Science can purify religion from error and superstition; religion can purify science from idolatry and false absolutes. Each can draw the other into a wider world, a world in which both can flourish.
—Pope John Paul II (1920–2005)

In my view, science and Buddhism share a search for the truth and for understanding reality.
—14th Dalai Lama (1935–)

Many who explore the origins and workings of our universe do so out of a sense of awe and wonder. Thousands of scientists from all cultures and beliefs are currently engaged in replicating the point at which matter appeared out of energy at the time of the big bang. CERN's Large Hadron Collider, housed within a massive circular tunnel, 27 kilometers in circumference, 175 meters beneath the border of France and Switzerland, represents not only the pinnacle of modern scientific engineering, it also carries with it the hopes of scientists young and old, that a deeper understanding of our origins will lead to a better quality and a more sustainable life for all. Their collective benevolent intent, their sharing of their enthusiasm and intellect, goes a long way towards reassuring me that this bold step into the unknown is both safe and necessary. These are scientists that are excited about what lies ahead, sharing astronomer Carl Sagan's vision that "Somewhere, something incredible is waiting to be known."

So it appears that faith is as important in science as it is in religion. My own search in the early 1980s for a deeper understanding of the science of healing led me into the then-strange world of Eastern medicine. Somewhere within me there was the trust and faith that such bizarre concepts as the Law of Five Elements, meridians, and chakras were valid, largely due to an acknowledgment that they had been born from the learned observations of wise and thoughtful humans, albeit from cultures very foreign to my own. I believe I have learned to apply and adapt these holistic principles effectively in my medical practice without decrying the value of safe modern medicine. Indeed, I have observed that Western and Eastern medicine draw great benefits from each other. For example, over recent years, an increasing number of elderly people with arthritis have been able to reduce their medication and increase the quality of their lives immensely by practicing the ancient art of tai chi. Many people of all ages are finding the practice of yoga an essential antidote to the stresses of busy urban living.

So over the years, my faith has been rewarded manyfold. The Law of Five Elements can now be explained in terms of self-similar fractal patterns occurring throughout nature. In figure 13, the outer circular ring represents the nurturing coherent relationships between elements, whereas the straight lines of the central pentagram show the controlling, possibly more chaotic actions of one element on another. For example, wood feeds a fire, but water may be needed to put it out if were to spread.

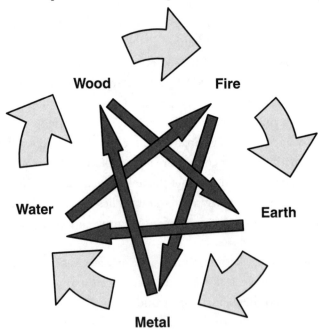

Figure 13. The five elements.

The Vedic seers observed the growth of human consciousness in terms of chakras, spiraling whirls connecting our inner and outer worlds. Opposing forces of control and submission need to be balanced within each chakra in turn, thereby allowing a steady rise in consciousness through the heart and beyond. This model is truly holographic as it allows for an instant nonlocal exchange of information between our bodies and society, and vice versa. We can clearly envisage the microscopic pattern of the DNA being projected holographically into macro form as the caduceus, the twin serpents of the "male" and "female" energy that rise as the kundalini, meeting in harmony at seven points in the body (see figure 14). Theoretically, with every chakra in balance, a state of coherence is achieved in the body, allowing us, as human antennas, to connect in perfect synchrony and in resonance with the universe.

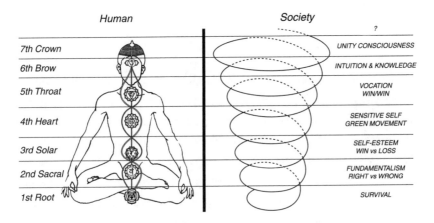

Figure 14. The spiral growth of human consciousness. (Adapted from the work of Don Beck, Chris Cowan, and Ken Wilber.)

The experiences we have here on Earth are the catalysts for this growth. Again, this is a two-way process, a mirroring—as our consciousness grows so does that of our society, and our universe. The cycle is completed as the field of consciousness is seen to act on our DNA, as previously described in chapter 10. This is illustrated in figure 15a.

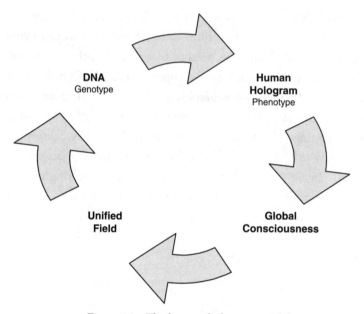

Figure 15a. The human hologram model.

This builds on, rather than completely negates, the traditional Mendelian/Darwinian model that depicts our genetic material as existing fixed within our DNA (our so-called genotype) in overall control of the growth and functioning of our body (our phenotype). Those with the strongest DNA survive and reproduce, hence we evolve through "natural selection." In addition, changes or mutations in the genetic structure occur over time in response to a changing environmental landscape—something the successful species has contributed to, and has become a part of (see figure 15b).

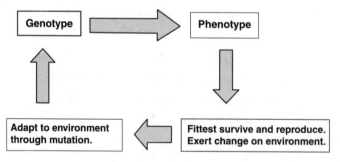

Figure 15b. The traditional evolution model—natural selection. Note: There is also now evidence that genes also transfer between species (horizontal gene transfer).[60]

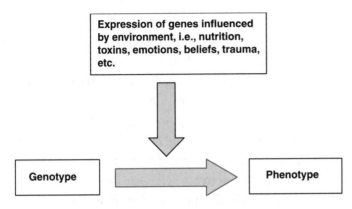

Figure 15c. The epigenetic model.

We now know that our existing genes can be activated, deactivated, and modified directly by environmental or epigenetic factors, including even the words and actions of those around us. Genes express themselves differently according to their immediate environment (see figure 15c).

One fascinating study illustrates just how powerful optimistic beliefs can be to the development of disease. In 2008, Dr. Robert Gramling and his team from the University of Rochester Medical Center analyzed data collected over fifteen years from 2,816 adults between the ages of thirty-five and seventy-five with no history of heart disease.[61] Remarkably, they discovered that men who believed they were at low risk to develop heart disease actually experienced a three times lower incidence of death from heart attacks and strokes. The research team accounted for, and factored in, such variables as family history, smoking, high blood pressure, and raised cholesterol.

This study suggests that fear itself, the belief that heart disease can develop, can have a dramatic effect on health. To quote Dr Gramling: "Perhaps we should work on changing behaviors by instilling more confidence in the capacity to prevent having a heart attack, rather than raising fears about having one." Of interest too was that in this study the presence or absence of this fear factor only seemed to relate to males.

The human hologram model, based as it is on physics rather than chemistry, delves more deeply into the process of epigenetics, theorizing on just how such beliefs come to exert their effects on our bodies. It looks beyond the chemistry of fear—the excessive secretion of cortisol, vasopressin, and adrenaline that, if unchecked, contributes to disease—to the science of fear itself. It even looks beyond the energy of optimism and fear—good and bad "vibes"—towards an understanding of how and why emotions, as pure information, pervade our consciousness. It acknowledges that information and consciousness lie at the very foundation of our being.

More and more researchers are acknowledging the role that compassion and altruism play in our advancement as a species. Psychologist Dacher Keltner is director of the Social Interaction Laboratory at the University of California at Berkeley. His research reveals that humans have evolved with "remarkable tendencies toward kindness, play, generosity, reverence, and self-sacrifice, which are vital to the classic tasks of evolution—survival, gene replication, and smoothly functioning groups."[62]

So it is clear we are not evolving exclusively through the process popularly known as survival of the fittest or "the law of the jungle." We are beginning to appreciate the vital role that human compassion plays in our physical and conscious evolution; we are all, it seems, also the product of a "survival of the kindest." What's more, we are actively playing our part in this compassionate evolution, dispelling fear with every simple act of integrity we perform here and now. Within the human hologram model, there is no conflict between creation and evolution; at every given moment, we are the cocreators of our future.

Chapter 12: About Time

Time flies like an arrow; fruit flies like a banana.
—Groucho Marx (1890–1977), American comedian

Of the four dimensions encountered here on Earth, time is undoubtedly the least understood. Perhaps the wisest comment on the nature of time, aside from Groucho's quip, is the popular "Time exists to stop everything happening at once." Indeed, if we are here to play our part in the evolution of consciousness, we need the structure of time and space to achieve our goals. We learn to plan, to save the effort of laboring over nine stitches when one would have sufficed if applied at the right time. There is little point in shedding tears when milk has been spilt over the leather upholstery of our brand-new car; there is every need mop it up thoroughly before the stale smell sets in, lingering for weeks on end.

Time, or at least our perception of time, is relative. Compare and contrast spending half an hour surfing the Internet with half an hour soaking luxuriously in a warm bath. The faster we go, the higher the broadband speed, the less time we seem to have to complete our search. Yet in wallowing aimlessly in the bath, thirty minutes can feel like an eternity. Time of quality, rather than quantity.

Most of the people seeking my help have what is termed chronic ill health. These are illnesses whose severity cannot be measured solely by the suffering encountered at any given moment. Someone who carries a diagnosis of cancer carries with it the worries and fears about the future. Will I die earlier than I had imagined? Will I suffer? How will my family cope without me? Will the chemotherapy make me feel unbearably sick? New variables are added to already complicated lives; plans are disrupted and priorities reset.

There were two gods of time in ancient Greek mythology: Chronos and Kairos. From Chronos, we derive the words "chronic," "chronicle," and "chronological." Chronos, like old Father Time, personified not only the wisdom, but also the physical decay that comes with age. Imprinted on my memory from a small boy is the weather vane depicting Father Time standing high above the grandstand at Lord's cricket ground in London. Here was the silhouette of a stooped and bearded old man, supporting himself with a rickety cane, while carrying a sickle across his back and an hourglass. The sickle, the symbol of the Grim Reaper, represents the inevitable harvest of human life after our time here on Earth.[63]

The image of the other Greek god of time couldn't be more different. Rather than personifying age and decay, Kairos was young, upright, fit, and active. He represented the fleeting moment, the opportunity to seize the day, or to quote the Roman poet Horace, *carpe diem.* Kairos was the joy and the lightness of being in the now.

For those suffering from the doubts and dreariness of chronic illness, learning to "live in the now" becomes essential to their healing. Almost always it is a relearning, as this is a state of being, a playfulness, that came naturally to us as children. It forms the basis of the simple exercises I describe in appendix 1—antidotes that can be performed instantly without draining precious energy. For it is within the moment, outside of linear time, that we connect with eternity. And with that part of us that never grows old.

Although our physical bodies follow the laws of entropy and weaken with advancing years, close relationships strengthen. Chinese medicine focuses on the cyclical patterns of nurturing relationships that occur in the natural world. The Law of Five Elements is also the Law of Five Phases acknowledging that our physical world and time itself are inseparable. When we gaze out at the stars, we also look back in time. A section cut through the trunk of an old tree reveals rings within rings, each one a living record of a specific time gone by. On the door of my rapidly growing son's bedroom, there are many small parallel pencil lines, each one representing a new height with a different date scribbled alongside.

We talk in biology of life cycles. Even with death, the cycle continues as our elements return to the earth, and our legacy—our relationship with the world we leave behind—lives on. Our days depend on the spinning of Earth in relation to the sun, our months, and with them the preparing of the womb for conception, on our orbiting moon. Our years are marked by the cycle of seasons.

These patterns of time are fractal—self-similar but with variations. From day to day, our weather is unpredictable, but our summers are always warmer than our winters. There is chaos within a pattern of order and symmetry. Figure 16 shows how the Law of Five Phases reveals itself in nature, our seasons, our lives, the creation of our universe, and our creative pursuits here on Earth.

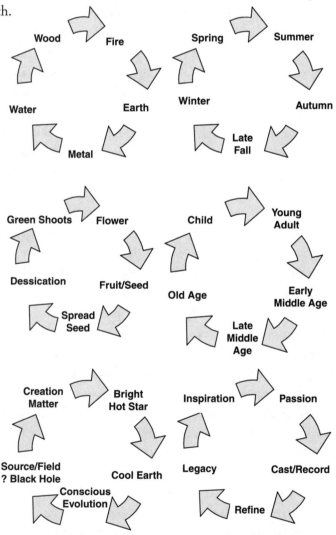

Figure 16. The five phases. Top left: five elements. Top right: seasons.
Middle left: plant life cycle plant. Middle right: human life cycle.
Bottom left: universe life cycle. Bottom right: creative cycle.

Notice how we "bloom" in early adulthood after "shooting up" in adolescence. How our dreams "come to fruition" through applying our minds to our hearts as we mature. How we eventually let go, passing on our eternal legacy to future generations.

Witness the vital role played by parents in our human life cycle. It may, at first glance, appear far-fetched to claim that growing a tomato plant from seed precisely mirrors the highly complex and responsible act of raising a child to adulthood, but bear with me. This spring I planted the tiny tomato seeds in a small plastic tray partitioned into tiny boxes or "wombs," each filled with the finest soil with the finest nutrients. I kept the tray warm on a window shelf in the garage, remembering—and sometimes forgetting—to water the seeds daily. Once the shoots had reached about four inches tall, it was time to send them into the world, replanting each in a larger container of fertilized soil. Exposed to the elements, pests, and our two cats, they still needed vigilance, protection, and watering, but over the next month my adolescent tomatoes grew rapidly. Their growth became wayward at times, so their stems required tying to the trellis now and then—just enough to rein them in, but not enough to cramp their style. Eventually, in full adulthood, they repaid me (and unfortunately, our friendly possum Gary too) with their succulent fruits, which were, according to my family, almost as good as "bought ones."

Nurturing our children into adulthood has proven health benefits, and we will explore this in section three. Our role as a healer mirrors our role as a parent. First, we engage empathetically as a protective guide to the person seeking our help. As the process matures, with hearts open, we share minds and responsibility, eventually releasing our protection until the time it may be sought again. When experiencing such deep healing, we often revisit our childhoods, coming to terms with the circumstances surrounding any abuse or abandonment. By understanding the motives of the abusers or neglecters, forgiveness for them and ourselves can ensue. Traumatic events in childhood, or adverse childhood experiences (ACEs), are now known to be significant contributors to the development of chronic illnesses later in life.[4]

For my own work, this is why the human hologram model is so important. An understanding that these cycles of time are self-repeating, truly fractal, gives us access to healing the wounds of the past, through the actions of the present. As the original traumas were more often emotional than chemical, then the healing too must be based on emotional truths and compassionate intent.

The model of the chakras can also help us heal past traumas. As we grow, we encounter challenges and lessons that lead to each chakra achieving balance in turn. In general, the teachings suggest that one cycle involving all seven chakras is completed by the time we are about thirty. Then a new cycle begins. So our next thirty years allow us to readdress the problems and traumas of the past; often we are shown the way by the actions and behavior of our children and of ourselves as parents. At sixty, another cycle starts as we learn to be grandparents. (See figure 17.)

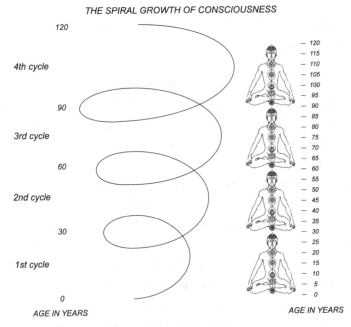

THE SPIRAL GROWTH OF CONSCIOUSNESS

Figure 17. The spiral growth of consciousness.

In this diagram, time is depicted not only in linear form, but also by cycles of consciousness opening and expanding while spiraling upwards. This model carries with it much optimism, by lending value to the wisdom of our elders, and by recognizing that there are many chances to heal the patterns set in the past. By healing the traumas of the past in this way, they are truly consigned to the past, and repetitive damage to our lives and to the lives of those who follow can be eliminated.

Understanding the fractal nature of time also raises profound questions about our own origins, and the origins of our universe. In Greek mythology, the gods, including Chronos, emerged from a primordial dark void known as Chaos. Modern science is now coming to terms with mounting evidence that all the galaxies in our universe, and hence we ourselves, emerged from the primordial dark voids we have come to know as black holes. In 2009, astronomers at the U.S. National Radio Astronomy Observatory (NRAO) publicized their conclusions supporting this theory after studying the formation of galaxies when the universe was still young, under a billion years old.[64] Some scientists would go further. Taking the holographic universe model to its ultimate logical limits, the very fabric of space-time consists of an infinite number of portals to other dimensions—John Wheeler's vision of a quantum foam.

The nature of time still remains perhaps our greatest mystery. Maybe it is part of the virtual reality we observe and create; as Albert Einstein remarked, "Time and space are modes by which we think, not conditions in which we live." An illusion that fades when we sleep, disappears temporarily when we are given a general anaesthetic, and more permanently when our "time comes."

However much contrived, our fourth dimension complements our first three perfectly. It remains the perfect device to help us achieve a life of fulfilment. And within the narrow limits of our present understanding, perceiving time as being relative, fractal, and cyclical, rather than absolute, linear, and finite, brings us a step closer to unraveling life's mysteries. At the very least, it is making us ponder from where, in heaven's name, we emerged. For it follows that this is where, eventually, we shall return.

Tenth Principle of the Human Hologram: Viable working models of the human hologram are emerging based on an understanding of the primacy of information and consciousness. Traditional philosophies such as those from India and China are based on this primacy of consciousness. Our four-dimensional space-time existence is seen as a consequence of other hidden dimensions, which, in turn, are vitally enhanced by its presence.

Summary of the Science of the Human Hologram

In 1971, Dennis Gabor was awarded the Nobel Prize for his pioneering work in creating three-dimensional holographic images from beams of laser light crossing to form interference patterns. The creation of a ghost-like three-dimensional object that appeared to remain in a fixed position when viewed from different angles has become a phenomenon in the worlds of art and communication. However, the science of the hologram has even greater implications for our deeper understanding of ourselves and our universe.

The notion that we live in a virtual reality as holograms in a holographic universe has gained some scientific and mathematical support over the past decade. By studying fluctuations in the energy patterns of outer space, researchers have theorized that our universe consists primarily and fundamentally of fields of information. It is through the power of our observation, through our senses, that this information manifests itself as the four dimensions we experience here on Earth. Logically then, we become cocreators of our reality, a reality that includes our very own presence.

For this paradigm to gain credence, we must first unearth hard scientific evidence that lends support to this revolutionary theory.

First, we must seek evidence that human beings do indeed influence outcomes of scientific experiments by the act of their observation. The twin slit experiments, with their many refinements over the years, have led the way for this phenomenon to become largely (but not universally) accepted within mainstream science.

Second, we must seek evidence that there exist other realms more fundamental to the materialistic world we encounter here on Earth. In the latter half of the twentieth century, physicists showed initially in theory, and then in the laboratory, that such realms were indeed fundamental to our reality—and the scientific age of quantum mechanics was born. However, science had yet to demonstrate that an essential feature of quantum mechanics, quantum entanglement (a state of instant togetherness independent of time and space), could exist outside the laboratory and within living beings. The technological advances in the early twenty-first century, especially in the field of nanotechnology, have now shown that quantum entanglement does indeed occur as a vital part of living processes. Sponsorship for this costly and sophisticated research has come from, amongst others, the computer, perfume, natural therapy, and solar-power industries. There is a need for more research into the pure science of quantum biology so that deeper insights into the human condition might be gained.

Third, the scientific study of the collective behavior of insects in their colonies and bacteria in a biofield is leading us to better understand the processes that allow complex multicellular organisms such as ourselves to function internally in a coordinated and synchronized fashion. Formal studies that prove a state of quantum entanglement, or a morphic resonance, in insect and bacterial colonies, have yet to be performed.

Fourth, fractal geometry has now become fully integrated into mainstream science. This has been made possible by the rapid advances in computer technology, the undoubted genius of Benoit Mandelbrot. Fractal geometry—the geometry of the natural world—reveals universal self-replicating sequences and patterns. These, in turn, lend insight into both the state of order and the state of chaos in nature. Man-made holograms display certain fractal qualities, with the information of the whole stored in the smallest part of the photographic plate.

Finally, revolutionary research by biophysicists Fritz-Albert Popp and Peter Gariaev has contributed to speculation that our bodies are formed continuously on invisible fields of information. Our DNA, it is theorized, is involved in projecting and processing a four-dimensional holographic map upon which our physical structure (as observed by us) forms. Our cytoskeleton, which incorporates such subcellular structures as microtubules, is further theorized to be involved in both local and nonlocal processes coordinating our growth and functioning. Our bodies are visualized as fully integrated complex quantum computers, with structures such as DNA and microtubules acting as foundries within which quantum states of probability "collapse down" into the recognizable realms of daily existence—our four-dimensional space-time.

All scientists now agree that behind our world of time and space lies another hidden from our senses. The science of quantum biology is still in its infancy. With the right will and funding, further research will lead to a deeper understanding of the human condition, our consciousness, and our unique role in the survival of our planet.

Ten Guiding Principles of the Human Hologram

First Principle: The study of holographic science and philosophy gives us a deeper understanding of the value of our everyday lives.

Second Principle: Holographic theory builds on ancient wisdom and modern science at a critical time in our history when information is being shared freely around the world, breaking down old barriers of race, religion, and politics.

Third Principle: The man-made hologram has two prominent properties. First, that of real parallax where the image appears to remain in a fixed position when viewed from different angles. And second, it has a fractal nature whereby all parts of the whole are contained in the smallest part.

Fourth Principle: Modern scientific and mathematical theory lends support to the theory that our universe is holographic. It follows that we, as part of the universe, must also be holographic. Modern science acknowledges that pure information is fundamental to our universe.

Fifth Principle: The science of the human hologram must include an awareness and study of our own participatory role as observers of ourselves, our world, and our universe. In observer physics, we become aware there is an ever-present subtle partnership between light, matter, and ourselves.

Sixth Principle: The observer effects we encounter at the micro quantum level do not always help us find that perfect macro parking space.

Seventh Principle: Our scientific understanding of entanglement at the quantum level of the very small is leading to an understanding of "entanglement" between humans resonating through compassion with each other, and with nature. Objective deductive science is complemented and enhanced by honest subjective human experience. Entanglement and the related terms nonlocality, bilocality, and multilocality are consistent with fractal nature of our universe and the human hologram model.

Eighth Principle: In the early twenty-first century, scientists are only just beginning to explore the quantum foundations of life.

Ninth Principle: The fractal nature of our universe, consistent with holographic theory, is now a recognized tenet of mainstream science.

Tenth Principle: Viable working models of the human hologram are emerging based on an understanding of the primacy of information and consciousness. Traditional philosophies such as those from India and China are based on this primacy of consciousness. Our four-dimensional space-time existence is seen as a consequence of other hidden dimensions, which, in turn, are vitally enhanced by its presence.

Section Three
The Human Hologram—The Experience

Chapter 13: Introduction to the Experience

Only one absolute certainty is possible to man, namely that at any given moment the feeling which he has exists.
—Thomas Huxley (1825–1895), British biologist

Every working day for thirty-five years, I have been asked to interpret people's feelings, and somehow make sense of them on their behalf. I have been trained to call them *symptoms,* a word derived from the Greek *sumptoma,* a happening. These are the happenings, or feelings, we experience that suggest that something is wrong inside. They are also our entry point to the medical system of every country on the planet.

As doctors, we are trained to listen to someone's subjective story, or history, at the very beginning of any assessment. Taking a history continues to be deemed in medicine to be the most important and fruitful part of our work. In the early '70s, as a medical student at Middlesex Hospital in London, much of my time was spent clerking new patients, meticulously recording their personal interpretation of their symptoms, before firing at them a series of detailed, and I'm sure rather annoying, direct questions. Only then did I attempt to examine the patient, record my findings, and nervously report them to my superiors.

So it could be said that medicine has always honored the prime importance of subjective feelings. After I qualified and progressed through hospital internship to more senior levels, this order of assessment became even more ingrained, although the pressure of time always bore heavily on the process. As a house officer in a West London hospital, every week one other doctor and myself were confronted by a waiting room of seventy patients with various types of cancer, many in advanced stages. On one occasion, a patient died while waiting to see me.

Even in the early '80s, after I had started in general practice, there were doctors in the neighborhood seeing a hundred patients a day. Those were the days of quick consultations and a culture, both within our profession and in the public at large, of seeking "a pill for every ill." Consultations were built around this mind-set, with the avoidance of open questioning and a free exchange of information, which could lead away from the medical model of prescribing towards the social and relationship issues so often at the core of the problem. Instead, doctors and the public alike, were content to engage in a line of inquiry that sought to match each ailment with a specific medication.

This social conditioning remains in much of the public. In general practice, I am still frequently expected to deal with four or five unrelated health issues within a fifteen-minute consultation. Frequently too, there is an opening line such as "My friends say I should start Prozac" or "The Internet site says to ask you for a free sample of _____" (invariably a medication whose name rhymes with Niagara!).

My own general practice is now built around a minimum of thirty minutes per consultation, with patients educated to know how best to manage minor illnesses themselves. Even so, new patients frequently present their feelings to me in medicalized jargon, as in "I think I have a gastritis pain," rather than "I have this burning pain right here in the pit of my stomach after food." Often too, a new patient will, rather than chat about their feelings of pain and fatigue, expect me first and foremost to look at their X rays—vague and frequently dusty two-dimensional black-and-white negative images often taken weeks in the past, and often adding precious little insight to their current problem.

And so, if I have become expert in anything over the years, it is as a listener to people expressing their feelings that something is amiss in their lives. I have repeatedly learned that, if progress is slow in someone's healing, this is the place to which we must return, with me listening even more intently than before.

In this section of the book, you will learn how best to apply the human hologram model to everyday life in positive, practical ways. At all times, I attempt to balance the objective, the science, with the more subjective, the experiences of myself and those seeking my help. The holographic paradigm, in line with modern scientific theory, places feelings, information, and consciousness at the root of our being. I am sure that if we are to make major inroads into healing our most persistent chronic health problems, we must relearn this truth and explore safe, effective therapies that can be fully owned and used by those who suffer. We health professionals must do our best to help our patients or clients achieve this goal.

By relating to the human body and the human condition in nonmaterial, more spiritual terms, in no way do we undermine those whose job it is to attend to our physical, mechanical needs. Hips and knees will still need to be replaced by skilled surgeons; children with cancer will continue to need complex chemotherapy regimens and bone marrow transplants in an attempt to cure them. There will be a place in the future for stem-cell therapy, and even gene therapy.

But it is my hope that the new paradigm, with its focus on understanding the science of feelings, will reduce suffering and prevent the onset and severity of much the disease we see in the early twenty-first century. This will happen by challenging and changing conditioned mind-sets. To have empathy, we doctors and health professionals have to challenge these mind-sets, held not only in society at large, but deeply in ourselves. We must examine our own past history.

As previously stated, there is now common acceptance that adverse childhood experiences (ACEs) have the power to influence not only mental and social health in later years, but also physical health.[4, 65]

A 2009 review published in the *Journal of the American Medical Association* acknowledges that there is now a strong scientific consensus on this issue. To quote the authors: "These early experiences can affect adult health in two ways—either by cumulative damage over time or by the biological embedding of adversities during sensitive developmental periods."[3]

Traumas and stresses experienced early in life are memorized by the body, often resulting in repeated patterns of behavior that compound over the years as lives and relationships become more complex. Abuse early in life can set the scene for abusive relationships throughout life. Even in the absence of these cyclical patterns of destruction, however, abandonment, neglect, and abuse early in life can lead to physical and chemical changes that leave someone at risk of developing serious illness in early middle age. One study showed that thirty-two-year-old adults suffering from depression were more likely to have elevated levels of a chemical known as C–reactive protein if they had a history of childhood mistreatment.[66] Raised levels of this protein are associated with an increased risk of heart disease and other inflammatory conditions.

We now know that the genes encoded in our DNA are neither static nor fixed. Genes for many illnesses can lie dormant and inactive within us, given the right conditions. Conversely, insults early in life, when we are at our most vulnerable, can lead to a "turning on" or an expression of these genes. An increased understanding of the processes involved has evolved into a whole new branch of science: epigenetics. Emotions and feelings often lie at the root of these epigenetic changes. So it follows that it is within the science of emotion and feeling that we should seek solutions.

Sometimes the effects of abuse in childhood can be devastating. I am grateful to Vanessa for allowing me to tell her story here. It is important to note that the presentation of type 1 insulin-dependent diabetes in childhood does not, in the majority of families, relate in any way to prior neglect or abuse. I know of many children, loved and nurtured by their parents and family, who have developed this very demanding condition.

Vanessa's Story

Vanessa grew up on a farm on the South Island of New Zealand. Her father worked hard on the land and often ended each day by drinking beer with friends at the local bar. Most of the time, he was strict but fair. There were times, however, when he returned home from drinking and became verbally abusive to Vanessa's mother. He occasionally took the strap to Vanessa's two older brothers, but never to her.

Vanessa's mother's life revolved around her home and her children. Trained as a secondary school teacher, she helped with the farm management and accounts during the day. At night, she cooked, and coped as best she could with her husband's controlling behavior. As far as Vanessa knew, her father was never physically abusive to her mother.

Her father's older brother also owned a farm in the area. He had two sons, both much older than Vanessa. She has vivid memories, beginning when she was four years old, of her cousins coming to stay for weekends, helping out on the farm and generally hanging out with her brothers. From the age of four to about eight, she was regularly sexually abused by one cousin, until he left the area at the age of eighteen to join the army. She never told her father, mother, or brothers of her ordeal.

At seventeen, during her last year at school (she was a bright student), she became ill. She lost ten pounds of weight over two weeks, and became very thin and tired. Her doctor diagnosed type 1 diabetes mellitus, and she was sent to the hospital for a week, where she learned to give herself insulin injections. Later that year, she met a local boy, a farmworker, and two years later they married.

The marriage turned out to be a disaster. He drank heavily and was both physically and verbally abusive to Vanessa. The level of glucose in her blood rose with the stress, and her diabetes became difficult to control. She became pregnant, and her diabetes worsened. She left her husband after the baby was born, moving to the city to live in a cold, damp flat, coping as best she could on a social welfare benefit.

Her medical condition worsened. The diabetes had affected her kidneys, which were beginning to fail. She had frequent trips to the hospital, eventually becoming dependent on a dialysis machine. She raised her young daughter to the best of her ability, helped on occasions by her mother who had by then returned to teaching full-time.

By the age of thirty, Vanessa's condition had deteriorated further. The only option was to consider a combined kidney and pancreas transplant, major surgery requiring strong medication to prevent the body from rejecting the donated organs. The surgery was successful, but two months later, a truck hit her while she was walking on a pedestrian crossing, fracturing her skull, both legs, one arm, and several ribs and rupturing her spleen.

Over the past five years, Vanessa has made a slow and brave recovery. She has to take more than fifteen medications a day, for pain, for sleep, and to prevent rejection of her organs. The powerful drugs she received have affected her immune system; she has developed multiple skin cancers that need constant treatment and surgery. Despite this, Vanessa is positive about the future. She has enrolled as an extramural student at the university, studying accounting, while looking after her twelve-year-old daughter, who is her pride and joy. Vanessa receives a disability benefit, while legitimately supplementing her income by helping a small local business with their accounts.

Vanessa is an intelligent and insightful woman. She knows that her life has been so complex that it would be simplistic to say that her poor health was solely due to the abuse suffered at the hands of her cousin and husband. To do so would place her in the role of a helpless victim, something she has continually striven to avoid.

However, over the years I have known her, she has developed the inner strength to recount more of her life story, and the more she has realized how much she has had to bottle up inside her. She knows in her heart that the control exerted on her life and spirit when she was at her most vulnerable has played a major part in her dramatic health problems. She has expressed to me how she regrets that she didn't receive in-depth compassionate counseling when young, as this, she feels, could have helped prevent the downward spiral in her health. She laments having to take so many medications now, spending her life dependent on doctor and hospital visits.

Prevention is better than cure. Her greatest wish, and this is something she is surely achieving, is for her daughter to grow up safe and well. And she knows her own immune system needs to be strong again and is determined to reach a place of understanding and forgiveness that will allow ongoing healing.

Listening to someone's story with a trusting and nonjudgmental ear is probably the single most powerful act of healing we can perform. As was the case with Vanessa, such trust can take years to cultivate. The listening must go on. The words must flow from the heart to be heard in confidence. And two hearts must be open, resonating in harmony with each other.

The scientific theories we have examined so far go some way towards explaining just how the apparently passive act of listening to words can be so healing. Words, when we are talking freely, seem to come from our lips in entangled, rather linear sequences. Patterns emerge, connections are made, and "the penny drops." Our bodies, not just our ears, hear the words. It may be that our DNA responds directly to these sequences, processing and relaying the messages holographically to the body as a whole. Maybe new shadows are projected onto the map, or matrix, of the body, creating an improved template onto which new resilient proteins form. And maybe this peaceful process can spread beyond our skin to our loved ones, and beyond.

Figure 18 shows how we strengthen the field of consciousness every time we listen with compassion. I am sure that if Vanessa's story has touched you in a meaningful way, then you have already played a part in her ongoing healing.

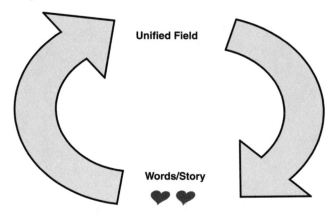

Figure 18. Compassionate listening.

Chapter 14:
The Heart of the Human Hologram

The thousand mysteries around us would not trouble but interest us, if only we had cheerful, healthy hearts.
— Friedrich Nietzsche (1844–1900), German philosopher

Your heart is the energy center of your body. Its beating pattern remains the most sensitive measurement of your state of relaxation, or anxiety. The average human heart beats more than one hundred thousand times a day, and more than three billion times in a life of over eighty years. A strong regular pulse is a sign of a healthy heart. A healthy heart displays, however, subtle variations in its rhythm that are undetectable through the traditional methods of palpating the artery in the wrist or listening through a stethoscope. To avoid undue wear and tear, the heart continually varies its rate and quality of beat very slightly from beat to beat. In recent years, computer technology has revealed that these very subtle variations of beat follow the rules of fractals.[67]

An alteration of this sequence of beat-to-beat variation is a vital sign of distress in a baby about to be born. As such, the baby communicates danger to the outside world directly through its heart. Studies show that adults who exhibit naturally enhanced levels of heart rate variability (HRV) are better able to withstand the stress of temptation. In 2007, research conducted by University of Kentucky psychologists Suzanne Segerstrom and Lise Solberg Nes showed this to be the case when participants with high HRVs in their study were more successful in resisting mouth-watering snacks. Instead, they showed tremendous self-discipline by chewing on carrots.[68]

Healthy hearts have a high level of HRV, a measure of an ideal balance of the parasympathetic and sympathetic nervous systems (or in Eastern terms, yin and yang).

Experienced practitioners of yoga have been shown to have higher levels of HRV, and are therefore thought less likely to succumb to heart disease.[69]

Yoga, with practice, induces a state of peace and calm, yet also resilience and discipline. A person with a healthy HRV tends to respond appropriately to threats by switching smoothly into a "fight or flight" (or sympathetic/yang) mode whenever necessary. When the threat has passed, the person returns swiftly to their existing state of grace and acceptance (a parasympathetic/yin mode.)

Another term for this harmonious parasympathetic state is coherence. Our heart does not simply pump blood to our tissues; it pulsates waves of energy. It is the drum to which our body dances. A coherent heartbeat creates a state of synchrony in our body like a drummer in a band, a DJ on a dance floor, or a conductor of a symphony orchestra. Its rhythm is the rhythm of life.

A coherent heart induces, or entrains, a state of coherence in the body. If we were to drop a pebble into a still pond, the waves would spread perfectly in ever-expanding rings away from the point of entry. If we were accurate enough to drop another pebble on exactly the same spot using the same force, repeating this over and over again at precisely the same intervals, we would see this pattern of perfect ripples continue.

When the ripples reached the bank, we would witness them rebound and collide with the outgoing waves. If we had a perfectly round pond with perfect banks, the rebounding and outgoing waves would eventually reach a state of harmony with each other, interfering coherently. There would be a pleasing symmetrical pattern of waves and counter-waves on our pond, as a state of balance or equilibrium were reached. If however, after the first pebble, we were to throw more pebbles at completely random rates and with differing force, the surface of the water would very quickly become disorganized and choppy. Waves would interact with others in a messy, chaotic way, without coherence or harmony. We can apply this wave model to the workings of the human heart.

One research group, the Institute of HeartMath, has focused on the ability of the heart to create, and reflect, a healthy coherence in the body. In a state of heart coherence, we are more receptive, open, and relaxed. We feel at peace. A coherent heart calms our mind, synchronizing with, or entraining, our brain waves. Similarly, by calming the busy-ness of our mind, we can enhance this state as head and heart achieve a harmonious balance.

If one measures on a graph the HRV of someone who is in a state of agitation, we see a jagged sawtooth "incoherent" pattern accurately reflecting someone who is "on edge." Compare this with the coherent waves of someone relaxed and receptive, going with the flow and "in the zone." (See figure 19.)

Stress/Agitation   **Peace/Harmony**

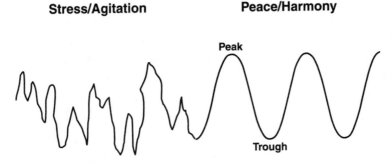

Figure 19. Heart rate variability. (After Institute of HeartMath Research.)

It is in this coherent heart mode that our physiology, our healing, is at its most balanced. And by achieving this state, we are able to influence those around us. It is the state I try to achieve in myself for at least part of each consultation. And it is the state I endeavour to facilitate in the person seeking my help, with the intent of giving that person the tools to achieve this herself. First, we must engage our hearts, then, and of equal importance, our minds.

The HeartMath scientists have devised practical "heart-focused" ways to achieve this state of coherence. As a therapist, I have found these particularly useful when adopting the listening mode, even on days that present many distractions. I focus on the area of my heart while breathing gently into my abdomen, which I allow to expand gently on the in-breath. I smile while looking into the eyes of the person seeking my help.

In a 1999 congress on stress, HeartMath researchers Rollin McGraty and Mike Anderson, and professor emeritus of Materials Science and Engineering at Stanford University William Tiller presented evidence that one person's coherent heart state can induce, or entrain, a relaxed (alpha) brain wave pattern in another person sitting a few feet away. However, it was important that this person too was in a coherent heart state.[70] The authors theorized that this may be due, at least in part, to the effect of one person's heart-generated electromagnetic field on another. The same authors previously found that a similar transference occurred when one person, in a genuinely caring state, touched another.[71]

Much of this pioneering research was performed in the 1990s and failed to be published in the mainstream medical literature. In recent years, however, the link between heart health and emotional states has become fully recognized, with many papers confirming this in major peer-reviewed journals. For example, there is mounting evidence that negative emotional states such as depression and inner hostility have direct links to heart disease. A combination of these two states is now regarded, alongside smoking, high blood pressure, and high cholesterol, as a major risk factor for heart disease. The levels in the blood of C-reactive protein, an inflammatory marker associated with ischemic heart disease, have been shown to be significantly higher in people displaying these traits.[72] Even persistent levels of gnawing anxiety such as ongoing doubts and insecurities have been shown to be associated with an increased risk of having a heart attack.[73]

The roots of heart disease are complex. It is clear, however, that hearts and minds that are conditioned from an early age to be on edge, guarding against constant threats, are at risk of disease later in life, unless balanced by a compassion that somehow has to be generated within. Mending such broken hearts often proves difficult, but the awareness that comes to someone on telling his or her story to another, whose heart is open, can be a vital first step on the road to recovery.

Ursula's Story

Ursula was born in Munich in 1958. The oldest of four children, her family moved to the countryside when she was six, as her father had gained a senior government position in farm management. He'd had a difficult and traumatic time in World War II. An officer on the eastern front, he was captured by the Russians in 1945 and held as a prisoner of war until his release three years after the war. He never talked of his combat experiences, but deaths on the eastern front numbered about thirty million (mostly civilians), nearly half the total loss of life for the whole of the war.

Ursula described her mother as being emotionally distant. Ten years younger than her husband, she had had grown up in Munich to the sounds of allied bombs exploding around her. As the raids persisted, she moved to the country to be brought up by an elderly aunt, while her mother, Ursula's grandmother, stayed in Munich to work in a munitions factory.

Ursula's parents met four years before her birth. Her mother was her father's second wife; his first wife had died tragically in her twenties of tuberculosis. By the time they moved to the country, Ursula already had two younger brothers, and during the afternoons after school was left to fend for herself.

On one such afternoon, she was playing by herself in the small barn on her parents' land when a farm boy of about fifteen appeared at the open door with a rope. He proceeded to tie her hands and feet together, then looking her in the eye, said the following words: "Never forget, wherever you are, whatever you do, I'll be there watching you."

Ursula can't remember any physical touching, nor did the farm boy strike her. Yet those words remained with her, terrifying her at night for years to come. She never told her parents.

At the age of fifty, the words haunted her still. She came to me suffering from episodes of chest pain and palpitations, occurring in a random haphazard manner. She had moved to New Zealand five years before, having separated from her husband in Germany. There was a traumatic custody battle over her two children, and her eight-year-old youngest son was still with his father. Although she was a successful businesswoman in her own right, there was plenty to be agitated about and to cause her to have pain and palpitations.

Despite talking about her current problems, practicing relaxation exercises, and having acupuncture, the heart flutters continued. After an initial medical examination proved unremarkable, I referred her to the hospital for a full cardiology assessment, where a long-standing heart muscle abnormality, known as a cardiomyopathy, was detected. As occurs with many such cases, no cause was found, and no sign of the illness has been found in other family members.

Ursula returned to me after the hospital investigations. The doctors were keen for her to start on medications, and to consider the possibility of having a pacemaker inserted if her symptoms returned. They were concerned that if the condition progressed, she would become a candidate for heart transplantation.

It was then that she opened up to me about the incident in the barn forty-five years before, and the words that had haunted her ever since. I referred her to a local health practitioner experienced in helping people recover from posttraumatic stress disorder (PTSD). Under the practitioner's guidance, Ursula focused on this one traumatic incident from her childhood, together with her feeling of not being nurtured by her parents. Night and day since, she has used exercises that release attachment to that trauma:

"Even though I was threatened by those words years ago, I now truly respect and love myself."

"Even though I received no emotional support from my family all those years ago, I now nurture, love, and respect myself."

After three months, her symptoms abated, and she now feels calm and confident. She has resumed yoga classes. Time will tell whether this will prevent Ursula from needing medication or undergoing surgery for her heart condition. Intuitively, Ursula feels that the main cause has been isolated, and is being deleted from her body. In her own words: "I feel free like never before, a sense of being emotionally cleansed."

Although we are now becoming aware of just how and why emotional trauma can lead to heart disease, there is plenty of room for optimism. Recent peer-reviewed studies lend support to HeartMath's earlier theories, showing that pleasurable emotions such as joy, happiness, excitement, enthusiasm, and contentment can significantly reduce our likelihood to develop heart disease.[74] There is even evidence that watching funny movies can help our heart; cardiac blood flow increases by 50 percent when we laugh "heartily."[75]

It appears that a healthy heart is one that holds joy, yet can adapt swiftly to pressures and stresses of daily life. It remains light and unburdened, but can pump hard and fast if its owner is being chased by a bull. The Chinese would say that the polar opposites, yin and yang, are perfectly balanced. Western scientists would say that the parasympathetic nervous system is in balance with the sympathetic. The metaphysician would reflect that the heart is the seat of the soul, a soul that grows in response to the demands of our earthly existence.

To the specialist cardiologist, the heart is first and foremost a pump. To the student of mind-body medicine, the heart has a governing say in a body that reflects the subconscious mind. The student of quantum physics may draw parallels between the entanglement of tiny atoms and photons in a coherent state at temperatures close to absolute zero, with the synchronized entanglement in our bodies that accompanies a coherent heart. The student of the human hologram model will view the heart at the core of the human biofield, a biofield linked holographically to other fields of consciousness, all unified into one great field. In chapter 16, we will speculate on how our heart, connected nonlocally to the unified field, reflects subtle variations in information or consciousness, which manifest within us as *feelings*.

The Institute of HeartMath shares this broader vision of human heart consciousness. Through their website, they coordinate the Global Coherence Initiative, joining coherent hearts around the world in unison at specific times. The aim is for the synchronized heart-centered intent of thousands of people to create a benign shift in global consciousness "from instability and discord to bal¬ance, cooperation, and enduring peace." The focus so far has been at times of natural disasters and conflict.

The greater aim, however, is preventative: cocreating a harmonious and sustainable future for generations to come. In short, large numbers of openhearted human are praying for the survival of their planet.

The institute's theories have been supported by research conducted by Roger Nelson and his team from Princeton University. In his Global Consciousness Project, random event generators scattered at various locations around the world recorded shifts at the time of, and just before, the terrorist attack on the World Trade Center on September 11, 2001.[76] In addition, magnetometers on two National Oceanic and Atmospheric Administration (NOAA) space weather satellites showed significant changes in Earth's magnetic field at this time.

Subtle variations in the heart's magnetic field are now regarded as vital early signs of heart disease. Heart magnetometers, developed by teams of researchers comprising doctors, engineers, and quantum physicists, are now regarded as state-of-the-art screening devices for many heart conditions. It is expected that a handheld device able to be used through clothing by trained nurses will be available in the next three years.[77] So, along with the measurement of the fractal nature of the human heartbeat, mainstream medicine is starting to embrace the quantum properties of the body. It is now being realized that this is our fundamental state of being, and hence detection of problems in this realm must result in improved health outcomes and the more economic use of health-care resources.

The human hologram model embraces the quantum realities of our bodies, linking them holographically with the cosmos. In theory, the magnetic field of our bodies, generated in the main from our hearts, is linked locally and nonlocally to that of our earth, our sun, and the center of our galaxy. In 2008, astrophysicists from Germany and the United States described the black holes at the center of galaxies as behaving like hearts, rhythmically and "gently" pumping energy into outer space in the form of hot plasma.[78] This is now thought to have an important regulating effect on the growth of stars. To quote Alexis Finoguenov, of the Max Planck Institute for Extraterrestrial Physics in Germany: "Just like our hearts periodically pump our circulatory systems to keep us alive, black holes give galaxies a vital warm component. They are a careful creation of nature, allowing a galaxy to maintain a fragile equilibrium."

So perhaps it is not surprising that a heart in perfect balance has true cosmic connections. After all, it seems we come from the dust of stars that emerge out of the abyss of black holes. A coherent heart is imbued with the wisdom of the ages. Ursula sensed this as she started life anew, drawing on the deepest resource of all to banish a terrible fear implanted within her while a vulnerable child. In William Shakespeare's Measure for Measure, the novice nun Isabella tries convincing the lecherous and hypocritical judge Angelo to reverse his death sentence on her brother Claudio. Claudio's crime had been to make his girlfriend pregnant out of wedlock. "Go to your bosom," Isabella urges Angelo. "Knock there, and ask your heart what it doth know…"

This simple advice remains for me the most profound and potent I have yet encountered.

Chapter 15:
The Brain of the Human Hologram

"All the same," said the Scarecrow, "I shall ask for brains instead of a heart; for a fool would not know what to do with a heart if he had one."

"I shall take the heart," returned the Tin Woodman, "for brains do not make one happy, and happiness is the best thing in the world."

—L. Frank Baum (1856-1919), American author of
The Wonderful Wizard of Oz

I can empathize with both the Scarecrow and the Tin Woodman as they joined Dorothy on her journey of discovery down the yellow brick road. Without a brain, the feelings from our body and the images, sounds, and smells from the world around us would make no sense at all. We wouldn't have a clue as to how best to respond to these messages or how to act appropriately. We wouldn't be able to remember, reflect, or debate.

As my father-in-law's Alzheimer's disease advanced unremittingly over the passage of ten years, his mind made its unerring retreat back through the ages of man. The worst time for George was at the point of early awareness—the understanding that something was happening over which he had no control. No amount of problem solving would fix the inevitability of his decline. Next he became an exasperating adolescent, neither an adult nor a child, but without the hope of a redeeming future. Then a child again who could laugh and cry without guilt, but who needed child-minders to ensure he was fed, clothed, and kept safe. As many freedoms were lost, a precious few were rediscovered. It was fine again to want his ice cream and before his main course; even better, this time nobody would scold or restrain him as he feverishly scooped the contents of his bowl into, and all around, his mouth.

It was as if he was now allowed to discard the guilt and conventions imposed on him by adulthood, and to, at long last, grant himself one simple pleasure that he should never have been denied. But such self-reflection was unlikely.

By now George could not recognize his image in a mirror. He still smiled at our faces, but we suspected that this was purely a reflex response to our own smiles. He retreated further into a strange infancy where, unlike a young child, he was loved for what he had been, rather than for what he was now, or was yet to become.

After he died, we were able to make friends with him all over again. At his funeral, we celebrated his acts of kindness, his humor, his art, his children, his grandchildren, his role in the allied invasion of Normandy in 1944, and his MBE (Member of the British Empire), the award he had received personally from Queen Elizabeth II for a lifetime of service in her government.

It is often said that we can only fully appreciate something's value once it is lost.

George's illness was without doubt my most valuable lesson on the workings of the human brain—no learned textbook or brilliant lecture on neuroanatomy or neurophysiology has come close. George, while possessed of a fully functioning adult brain, was able to make a tangible difference in our world, playing his special part in the evolution of our species, and the growth of our consciousness. We may feel with our whole body, but we process these feelings in our brain, devising ways of putting them into action for the greater good.

Our brains are without doubt the most complex, and the least understood, organs in our bodies. Our brains are constantly struggling to understand what they are. More often than not, as the Tin Woodman would agree, racking our brains to find such answers rarely leads to a state of happiness, rather one of confusion and frustration.

And brainpower alone is insufficient to heal us from chronic complex illnesses; if the reverse were true, clever people would not get sick. (Even Hugh Laurie's TV character, the intellectually gifted but emotionally damaged, opiate-addicted Dr. Gregory House, would have to agree with this!)

In fact, we are all struggling to cope with the mass of information that has become available to us over the past decade. "Keep it simple, stupid,"

"Less is more," and "Way too much information" are the catchphrases of our times, as we try to prevent the dizziness and confusion of being caught up in the IT revolution. To find balance we are having to find effective ways to chill out, and to fit them into our ever-demanding schedules.

Yet there are times that all this becomes too difficult. Distressingly, our future doctors are more likely than any other students to become overwhelmed and depressed. Studies have shown that up to one-quarter of all first- and second-year medical students may suffer from varying degrees of clinical depression. In a 2005 review in the *New England Journal of Medicine*, the authors quote a fourth-year Harvard medical student who estimated that three-quarters of her close friends in medical school had taken psychiatric medications at some point during their four years of study.[79]

Much of my motivation in writing this book, and in making my career changes, has been to address the imbalances that threaten to make both doctors and patients sick. As with many of the doctors of today, I chose this path because of a heartfelt feeling to help others, whilst—I admit it freely— experiencing a fulfilling and comfortably rewarding life in the process.

The excessive amount of medical information a young doctor's brain has to process is, of course, only part of the problem. Medical students now carry a heavy burden of financial debt—student loans that either they or their parents will have to repay. They miss the sleep that their young bodies need. They confront death and the dying, expecting to cope in ways far beyond their years. They begin to feel squeezed into a narrow constricting tunnel that leaves little space to explore the deeper meaning of their lives, and the lives of their patients. So for both doctors and patients alike, it would seem the time is right for the emergence of a more complete, holistic model of the human body, and for an understanding of just where in this model our brain fits.

Unfortunately, our brain remains the organ about which we know the least. Even the owners of our most brilliant brains cannot explain to us how their prized organs function: "Our very awareness," physicist Stephen Hawking once remarked, "the awareness that constructs and analyzes fractals and everything else, continues to remain a mystery to itself."

And without doubt, our brain is a fractal organ, with its higher centers packed in tight in the shape of highly convoluted folds and ridges. This allows our skull to house a maximum number of cells in a very confined area, with highly efficient channels of communication between them. As with the Mandelbrot set, and the intricate coastline of an island nation, the further we venture into the peripheries of a brain, the more complex it becomes. To quote Ian Stewart, the renowned professor of mathematics at Warwick University: "There is a natural evolutionary route from universal mathematical patterns in the laws of physics to organs as complex as the brain."

For the purposes of this book, the enjoyment of its readers, and the sanity of its author, I intend to steer as clear as possible away from the complexities of neuroscience. In fact, the closer we look with our microscopes and computer-assisted scanners, the more confusing it is for all but those who specialize in this fascinating subject. So whilst fully acknowledging its importance, let's focus instead on approaches that broadly comply, in theory at least, with our model of the human hologram.

Reductionist neuroscience will surely play its part; however, I believe we will only derive a truly deep understanding of human consciousness by approaching the subject from many different angles. Linear thinking must join forces with lateral thinking; past experience with intuition; head with heart. We must meet the brain on its own terms.

Karl Pribram and the Holonomic Brain

It was Karl Pribram, the Austrian-born neurosurgeon and professor of psychiatry, who in the early 1990s, in collaboration with physicist David Bohm, mounted a serious challenge to the established view of the human brain. Pribram first became intrigued by reports of animal experiments performed by the biologist Karl Lashley in the 1920s. Lashley had shown that widespread cuts through the brains of animals did not eradicate learned behaviors produced by training. Pribram's curiosity was further piqued on learning of Dennis Gabor's groundbreaking work on holograms in the 1960s. He forwarded the theory that information could be spread holographically throughout vast expanses within the brain. He named this the *holonomic brain.*

Pribram's theory is still very much a work in progress. It relies heavily on his expert knowledge of the architecture of the brain and mathematical predictions. I'll present a simplified version here.

In essence, our brain adopts a role similar to that of Gabor's photographic plate. However, instead of the highly organized coherent beams of laser light that interact on the plate, our brain has to cope with far more complex, scrambled messages. Helping our brain are our senses, which Pribram describes as lenses transforming the magnetic waves of sound, light, smell, and our bodily sensations into a pattern containing bits of information within our brain (again, as in Gabor's flat photographic plate). This transformation is known as a Fourier transform. It can be demonstrated by observing how a lens spreads and separates light into the lightest and darkest components. As a child, I learned how to use a magnifying glass to focus the rays of the sun onto a point on a piece of paper. After a short while, smoke would rise from this point, and then the paper would spontaneously catch fire. What I didn't know then, nor would I have likely cared, was that the other "darker" waves of light were pushed aside from the central spot and scattered uniformly over the paper. Unbeknownst to me, I was witnessing my first Fourier transform.

A similar process is now thought to happen at the back of our eyes, which is why it is rather important not to stare directly at the sun. Light energy passes through the lens at the front of our eye, where it is refracted to the retina at the back.* The information that gathers here now exists in a form (a Fourier transform) that can be transmitted via the optic nerve and pathway to the visual cortex. Along the way, more information is added to the mix at a knee-shaped node, called the lateral geniculate body (LGB).

This new information comes from our other senses, and our body as a whole, so the information arriving at our visual cortex is a harmonious blend, helping us put our vision into the context of our space-time environment. According to the physicists, the mathematics strongly suggest that all this information must be spread through the visual cortex in the pattern of a Fourier transform. (See figure 20.)

---

*There is mounting evidence that quantum processes are at work in the retina of many animals, playing a vital role in navigation and orientation.[33, 34]

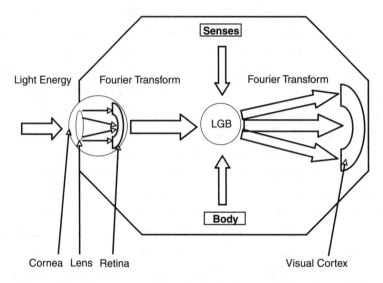

Figure 20. Light energy encoded in the brain. The model of our visual perception presented here is simplified. For example, our optic tracks cross, allowing us to have binocular vision; and modern scans suggest that not all the optical processing occurs just at the visual cortex.

Now comes the magic. Something, somehow, has to convert all these scattered bits of information into a picture of the world around us. And if, indeed, our holographic universe is really a blurred mess of interfering fields, the visual cortex is the site where much of our virtual reality is created. And the magic wand that makes all this possible? I suspect that if I were to say that the wand is our consciousness, it could provoke either derision, or a barrage of questions about exactly how, why, and possibly even who.

Before we consider some of the possible mechanisms behind the creation of our perception, some further clarification is needed around Pribram's proposed model. He theorizes that each sense processes its information in the way I have just described, at *specific* locations around the brain, that is, the visual, auditory cortex—a process, it could be argued, that is not truly holographic. So in this regard, our brain is functioning as a standard electronic computer, receiving and encoding data, then relaying the data back in a usable form. He also theorizes, however, that *each brain cell* at these specific locations can have access to *all the information of the whole,* just as some suggest each photon of light reaching our eyes has access to the whole. So we are now describing the sophisticated workings of something yet to be created by the world's most progressive IT companies: the quantum computer.

And considering our brain as a quantum computer is the necessary next step towards solving the riddle of just how our consciousness manifests itself. One prominent researcher from Florida State University, Professor Efstratios Manousakis, has forwarded the theory that quantum processes are at play when we are confronted by ambiguous images, such as the famous vase that can equally be perceived as being the profiles of two faces. The image exists in a state of probabilities, until, through the act of our perception, it collapses down into being either a vase or two faces.[80] (See figure 21.)

Figure 21. A vase or two faces?

And what, you may ask, are the specific mechanisms that lie behind our human consciousness? And where exactly in our brain does this happen? In an attempt to answer these questions, a special partnership has formed between the owners of two of the most searching minds in the Western world. Fittingly, one half of this duo is a practicing anaesthetist, a doctor highly trained to render folks unconscious, then conscious again. The other half is a gifted professor of mathematical physics. Their names are Stuart Hameroff and Sir Roger Penrose.

Microtubules and the Holographic Brain

Microtubules exist everywhere in our body. As their name suggests, their most common shape is tube-like, a triple helix of proteins molecules wound tightly around a central tunnel. They are a vital part of our cytoskeleton—the connective tissue matrix whose branches reach out in fractal form to every corner of our body, ending in tiny filaments and antenna in each of our fifty trillion cells. The cytoskeleton accounts for no less than 70 percent of the volume of each cell.

The cytoskeleton with its microtubules even exists in tiny organisms that have no brain or nervous system. The single-celled pond-dwelling paramecium that has dwelt on the planet for roughly 540 million years does, however, have a very sophisticated cytoskeleton. This web not only provides the scaffolding that helps it maintain its firm cigar-shaped structure, it also governs its growth and even every move it makes. The branching web ends in thousands of cilia or strands that beat furiously, propelling it through the water. Even without a brain or nervous system, the tiny paramecium maneuvers itself to and fro, avoiding objects and apparently making "decisions" about exactly where to go.

Microtubules are present in the cells of all animals and of the vast majority of plants. They are, however, absent from blue-green algae, bacteria, and viruses. In 1976, astronomer Carl Sagan speculated that over half a billion years ago, a time when only these primitive microtubule-lacking organisms existed, they became "infected" with organisms with cytoskeletons, possibly highly mobile little creatures with tails like spermatozoa. In other words there was a mating of two species, hitherto alien to each other.

Much of these observations, and underlying research, comes from Sir Roger Penrose's book *Shadows of the Mind,* published in 1994. This was the second of a trilogy of volumes expertly explaining these complex issues to the lay public. His underlying thesis is that the intelligence of life cannot be explained in the purely linear terms of classical physics. There is much more likely to be a highly sophisticated quantum element, at present preventing human beings recreating such intelligence artificially. This, of course, is hotly debated by the proponents of artificial intelligence (AI). However, scientists from both sides of this debate would have to agree on one salient point: that the most up-to-date computer-operated robot possesses an intelligence that pales in comparison to a common ant.

Together with Stuart Hameroff, an anaesthetist from the University of Arizona, Sir Roger Penrose developed the hypothesis that the microtubules in each cell play a part in our consciousness, that the microtubule is a site where the quantum world and the classical world of space-time meet. A site where the quantum world collapses down into the physical world—the site of our consciousness and our perception.

As a student, Hameroff was intrigued by watching the intricate process of cell division—something that to this day I myself view with curiosity and wonder. The spindles that pull the chromosomes apart in this tug-of-war—or more appropriately, "tug-of-life"—are fibers consisting of microtubules. Before performing this miraculous act, they escape from their central control center in the cell, migrating en masse towards strands of DNA in the nucleus, to which they attach themselves.

The act of cell division, or mitosis, then happens with the paired chromosomes being meticulously prized apart, a performance essential for the growth and reproduction of the organism. It was after reading Penrose's groundbreaking book *The Emperor's New Mind* in 1991 that Stuart Hameroff felt compelled to contact him to share his thoughts. As a practicing anaesthetist and medical researcher, he had already published many papers on the role that the microtubules within our brain could play in our waking consciousness. Since then, together they have further researched in great detail the microscopic structure, and possible function, of these microtubules—a process they call Orch-OR (orchestrated objective reduction).

As our brain cells only rarely divide and multiply, the microtubules in our brain must have functions beyond their vital role in cell division. They are certainly very prominent, and numerous, within our brain cells, often running the length of each neuron with their tips starting (or ending) in the dendrite portion close to the electrical joining points between cells (synapses). Figure 22 shows how the microtubules are arranged within a neuron in the visual cortex.

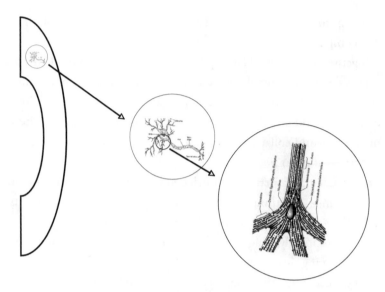

Figure 22. Visual cortex (left), neuron with fractal-shaped
dendrite (center), and rows of microtubules (right).

The fractal arrangement and structure of the proteins that make up the microtubules is important, as is how they combine to make larger structures. For example, encoded within their intricate spiraling patterns of proteins are classical Fibonacci series of numbers, linking them mathematically and holographically to so much inside and outside our bodies.

The proteins that form these intriguing patterns of molecules are known as *tubulins,* and exist in pairs with complementing, but opposing electrical charges. This in theory allows the microtubule to function as a classical computer that encodes information into a binary form, or "bits" of either 1 or 0. The efficient fractal packaging of the tubulin molecules, and possibly their "sacred" geometry, makes the microtubule a credible candidate for our cell's computer hardware. However, Penrose and Hameroff have envisaged a far more sophisticated, and efficient role, for these fascinating subcellular structures—no less than that of a *quantum computer.*

The quantum computer has yet to be manufactured commercially. When it inevitably makes its appearance, it is set to revolutionize the IT industry. It will possess a vastly increased storage capacity—instead of bits, its information will exist in another dimension (in a superposition state) as merely probabilities of 1 and 0, or qubits, entangled with countless other qubits in a state outside time and space. So, in essence, there will be absolutely no data storage problems!

The paradigm-stretching idea is that the microtubules act as both classical and quantum computers. Penrose and Hameroff describe how these states alternate rapidly and continuously, blending universal quantum information with classical information processed "unconsciously" by the brain from the body. They theorize that there is an elaborate feedback mechanism within the microtubules, orchestrating a harmonious balance between the quantum realm and the classical world into which it collapses down.

The main criticism of their model revolves around the very reason why the quantum computer remains presently so difficult to manufacture. Until recently, it was thought that entangled states could only exist near the incredibly cold absolute zero, as in Bose-Einstein condensates, or at the extreme heat that facilitates Fröhlich's condensates.[81] This argument is less convincing, however, since the recent demonstration of quantum processes occurring at room temperature in photosynthesis.[28]

The other main criticism is mathematically based. So far, *decoherence* or collapse of the quantum state has been demonstrated only to occur extremely fleetingly, and hence there is doubt that it can be sustained for long enough to play a role in our consciousness. Penrose and Hameroff counter this argument with mathematics that show that, indeed, it can. (More details on this in chapter 23.)

In broad principle, however, the theory that microtubules play a role in our consciousness, and perception, complies well with our human hologram model. It acknowledges that a reality exists "out there" dimmed from our daily gaze, that is a primal force—a matrix or field of information of which we are a living, feeling part. It complies closely too with Beckenstein's calculations that all the information in the universe could be stored on a massive flat screen in the tiniest of Planck scale pixels.

To quote Stuart Hameroff: "[T]hus the infinitesimally tiny Planck scale, described by loop quantum gravity, string theory, quantum foam etc., is the authentic matrix whose configurations give rise to conscious experience (and everything else)."

As the microtubules exist in such numbers in every brain cell, it is theorized that this "downloading of the matrix" is a universal feature of the brain as a whole. If we return to figure 22, we can see how prominent microtubules are within our visual cortex. We can speculate that this is where the final conversion of the visual information received from the back of our eye takes place. That this is where the Fourier transform, or encoded information, is magically converted into the shapes and colors we perceive of as our reality. The same process applies to all our other senses, including the sensations of touch coming from all parts of our body.*

The Penrose-Hameroff theory is an important step towards a deeper understanding of human consciousness. It necessarily focuses on one organ, the brain, the proponents' specialized area of expertise. Together, as a biologist and a mathematician, they form a unique and impressive partnership; their theory is all the more credible, and potent, for this.

As a general practitioner, I am required to know a little about a lot, a jack-of-all-trades and perhaps a master of only one—the study of relationships. Unlike specialists, general practitioners do not have the luxury of isolating one organ system from another; in fact, we are the people who do the sorting out, granting our specialist colleagues the freedom to focus unimpeded on the organs they have come to know and love.

With this in mind, it is now time to place our most complex and perplexing organ in its rightful place, encouraging its role as a helpful, wise relative rather than an overpowering patriarch to our other organs. In my experience, a state of coherence within the body as a whole is best achieved once the active, thinking mind is quieted. One of the simplest and easiest ways to achieve this state is by focusing away from our brains, to our breath. Alternatively, we can focus our attention on any other part of the body, ideally together with gentle breathing. In this quiet state, we become more aware of our feelings and, following the model just described, more connected to "the authentic matrix." So to quiet the mind, to become synchronized to the field, to achieve coherence, we must, paradoxically, escape from the brain and take refuge in our body.

Let us now examine our rational brain, and place it in some kind of perspective.

---

*Some neuroscientists now theorize that there also exists a state of quantum entanglement between our sense organs and the processing parts of our brain, allowing instant *teleporting* of quantum information between theses sites.[81b]

The Rational Mind

Over the last few pages, we have been immersed in the complexities of the human brain. Although the theories I have outlined have come from scientists regarded as more holistic than many, we have had to take a classical, reductionist approach to the subject of consciousness. We have had to think with a focused mind, largely ignoring the other messages coming into our brains and being. If we were to measure our brain waves with an electroencephalogram (EEG) while reading this, we would likely discover a pattern known as the beta state, a high-frequency state of 12 to 30 hertz (Hz) associated with mental concentration. We may even find evidence of the even higher frequencies of about 40 Hz, the gamma wave pattern thought to be associated with the very act of perception, particularly the process of visual perception described in this chapter.

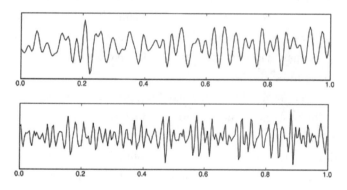

Figure 23. Beta wave pattern (top) and gamma wave pattern (bottom).

We have had to be in this mode to concentrate on something that is perhaps not easy to understand. In this mode, we are using our thinking brain in a state of wakefulness. We are, at the same time, making sense of new information, working out precisely where it fits into the world we know, the world we perceive of as our reality.

When neuroscientists and doctors use the word "consciousness," they are, more often than not, referring to this state of thinking wakefulness. However, others would feel that this is an altogether too limiting description of the word. To be fully conscious as a human, they would argue, one has to be aware that other realms and dimensions exist, and that our thinking mind is but one element of a much greater consciousness. The Greek philosopher Plato was firmly of this view; together with his mentor, Socrates, he was highly critical of those who directed their full attention to studying the complexities of the material world. "Entire ignorance is not so terrible or extreme an evil," said Plato, "and is far from being the greatest of all; too much cleverness and too much learning, accompanied with ill bringing-up, are far more fatal."

Our rational mind, of course, plays an essential part in our lives. Without humans' advanced ability to reason, there would be no technological advances, no keyhole surgery, no Internet, no blogging, and ultimately less chance for individuals to air and share their views.

The hard-earned rationality of an airline pilot, and her air crew, allows us to feel safe and relaxed while being propelled through space at 550 miles an hour with only a layer of metal separating us from Mother Earth 35,000 feet below. Moreover, our own rationality helps us every day with every task we perform, whether exciting or mundane. It recently allowed me to assemble a flat-pack bed in a little under six hours, something the young furniture salesman had reassured me would take his "grandmother fifteen minutes maximum."

But as yet, rationality alone has been unable to solve many of our world's major problems. There is no cure for most cancers; wars continue; bigotry persists.

Rationality requires training, and disciplined learning. By and large, our schools and universities have adopted a role of teaching us how to reason, and how to relay reasoned arguments to others. So we are conditioned from a young age to reason and become clever, but as Plato pointed out, if our upbringing was somehow "ill," this in itself could prove "fatal." We only have to bear witness to the Hitler youth movement, the Chinese Red Guard, the Jonestown massacre of last century, and the terror of fundamental extremism that plagues this century to understand that Plato was not overstating his case.

Albert Einstein, possessing arguably the most brilliant mind of the twentieth century, valued intuition above rationality. He is quoted as saying: "The intuitive mind is a sacred gift and the rational mind is a faithful servant. We have created a society that honors the servant, and has forgotten the gift."

The Intuitive Mind

During this chapter, your brain has patiently joined my brain in a journey of discovery into its own workings. The deeper we have delved, the smaller and more intricate the subjects of our examination have become. And yet the answers to the most important questions somehow remain elusive. We may journey right into the center of a tiny microtubule, or inside a DNA molecule, in an attempt to discover the holy grail of consciousness, only to find a space, a portal or door, that opens to a multitude of totally unfamiliar dimensions. The Chinese would say that at the heart or source point of any yang adventure, there is absolute yin. The closer we examine the machinery of our brains, the more materialistic our approach, the closer we come to realize that at the root of our fullness lies a point of emptiness.

The rational, reductionist path of investigation that preoccupies Western scientific thought is the necessary yang element to the yin of Eastern-based intuitive practice. In Eastern cultures (and in most ancient traditional cultures all over the world), consciousness, spirituality, and the nonphysical world are fundamental to their existence, and to the world to which they belong. As a rite of passage into the realities of the universe, a shaman in the Peruvian rain forest will guide young tribal members through an altered state of consciousness, facilitated by a natural psychedelic compound such as Ayahuasca. The experience is not intended as an entertainment, or some fun party piece; it serves as a confirmation that there exists far more than our physical dimensions of time and space. In traditional cultures, the young are nurtured through such an experience by their elders, with the collective support of their tribal family.

Our daily life requires us to have a persistently active, rational mind. This constant internal chatter needs to be somehow quieted for us to access a state of being wherein our intuition and creativity can flourish. In my experience, the safest way to achieve this is through the regular practice of *meditation,* and the simpler the method the better. My definition of meditation is broad: for those who are starting out, five minutes a day of turning their attention from the active thoughts in their heads to the simplicity of their breath is all that is required.

It is through meditation that we gain a far deeper understanding of the word *consciousness;* we become increasingly aware that being fully conscious as humans means more than simply not being asleep. Nor is this consciousness confined to our brain; to achieve the necessary state of balance, or coherence, our body needs to be at peace. It is by paying attention to our feelings, our "body mind," our intuition, that we develop this more profound understanding of the human condition. As Einstein said, this is the sacred gift to which our rational mind must become a faithful servant.

The Vedic seers were seasoned meditators. They envisaged the growth of human consciousness in the form of twin serpents—of the opposing and complementing forces of male and female energy—rising up through the body. At each point they crossed, perfect balance was reached, allowing a higher level to be processed and subsequently balanced. The forces were by and large competitive until they reached the heart—at the heart, perfect balance facilitated unconditional love for others, and oneself, leading to an ascension to higher realms of consciousness.

From the heart, we traverse upwards through the throat area where one's true vocation is realized; then on to the brow and crown—a brain that, in balance, can cope with the concept that each of us is completely unique, yet also connected to a universal consciousness.

Each step is known as a chakra or wheel, depicting a spiral whorl, or wormhole connecting our body holographically to universal nonlocal realms. So although true balance in our brow and crown chakras represents a state of enlightenment, connecting us with awareness to the unified field of consciousness, it can only come to pass once balance has been achieved at every step.

There is Western scientific evidence that heart-focused exercises, with their associated feelings of peace and acceptance, can lead to greater mental clarity. This was already covered in the previous chapter. In purely electrical terms, our hearts are far more powerful than our brains. The electrical potential of our hearts, as recorded on a standard electrocardiogram (ECG) is a thousand times greater than that of our brains, as recorded on a standard electroencephalogram (EEG).

When we are in a relaxed, intuitive, and creative state, an EEG will register brain waves in an alpha state, with a frequency approximately between 8 and 13 Hz.

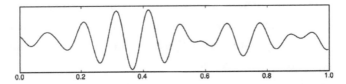

Figure 24. Alpha brain waves.

Interestingly, this pattern of brain waves only occurs if our eyes are closed. In a process known as alpha-blocking, the pattern is eliminated as soon as we open our eyes. Alpha-blocking even occurs when, with our eyes still closed, we turn our attention to the environment. Opening our eyes in a totally dark room, however, will not produce alpha-blocking.

This finding adds substance to the theories of Karl Pribram, Roger Penrose, and Stuart Hameroff, as outlined earlier in this chapter. To recap briefly, light energy is processed by our eye into information that can be transmitted to the visual cortex, and its microtubules, for further processing into our familiar dimensions of time and space. The electrical activity thought to be associated with this process in the cortex—the collapse of the matrix into our classical four-dimensional world—is at the much faster frequency of 40 Hz (gamma waves).

It is important at this point to understand both the strengths and limitations of the standard EEG, a tool that has proved immensely valuable in clinical medicine for over seventy years. On the upside, it provides an instant record of the electrical activity of the brain, and is reasonably cheap to perform. Unlike newer scanning methods such as functional MRI (fMRI), it measures electrical activity directly, rather than blood flow though parts of the brain.

On the downside, though, it only records information coming from the surface of the brain, its neocortex. The ten billion neurons are arranged here in columns, only six cells deep, at right angles to the skull. This highly organized architecture allows the electrical impulses to be transmitted effectively to the many electrodes placed on the scalp. However, much information is lost and canceled out on the way.

By scattering such a large number of detecting electrodes over the scalp, the EEG captures only a rough overall approximation of the real activity in the cortex; in truth, there is a much more chaotic electrical pattern, with considerable variation from zone to zone. And it misses out entirely on recording the activity of the vast majority of the brain that remains out of range, lying too deep for its sensors.

Despite these limitations, however, the simple EEG pattern can prove very useful in clinical practice. For example, those suffering from chronic alcoholism exhibit less alpha activity than others. As they undergo supervised programs of relaxation known as alpha-theta training, they witness a change in their EEG pattern to the normal alpha pattern, which, in turn, serves as powerful motivator for further healing.[82]

The oversimplified patterns of the standard EEG have been subjected to computer-based mathematical (Fourier) analysis in recent years in an attempt to identify and quantify more useful underlying patterns. Although these quantitative EEGs (QEEGs) are undoubtedly an important advance, they are also noisy and expensive. Many therapists find the results too complicated, perhaps too close to the true state of chaos reflected within our brains. However, QEEGs do provide us with more accurate measurements of brain activity, giving skilled therapists more options as they train their clients to suppress "abnormal" patterns, and to elicit "normal" ones.

QEEGs have also shown that alpha-blocking is not the black-and-white, absolute effect suggested by the standard EEG. In fact, the phenomenon has now been renamed alpha desynchronization, as different parts of the cortex can be switched on or off at any given time. And alpha rhythms can be shown to be partially blocked, for instance, when sensory information is first anticipated.[83]

But let's now return to simpler matters, thereby inducing more alpha activity in our own brains. In this state, we dissociate from our conscious connections with the material world, and return to a childlike state of wonder and imagination. There is less noise to distract us from our feelings, sometimes known as our subconscious.

We may even enter a light hypnotic trance, a state of mind altered from our waking, thinking, rational mode. We are also vulnerable, open to the powers of suggestion. At the snap of a stage hypnotist's fingers, we can become a clucking chicken, or even more embarrassingly, a teapot. As young children spend much of their time in this playful state of consciousness, they are vulnerable to the overpowering suggestions of others. Their perception of reality can be influenced for life, fears can be infused into their being, manifesting as ill health years later. More positively, happiness, love, and safety can also be processed by the child in an alpha state, reducing the risk of a sad, lonely addictive adulthood.

As we slip more deeply into our hypnotic or meditative trance, our standard EEG now shows a theta rhythm. The pattern has slowed, the frequency of the waves being 4 to 7 Hz.

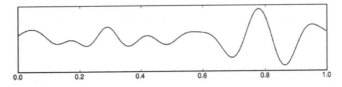

Figure 25. Theta brain waves.

In this state we are truly "in the zone," in balance with the Schumann resonances of nature. We are now deeply in touch with our subconscious, and may recall, with prompting, events in our childhood blocked from our everyday awareness by the busy-ness of our thinking mind. In this more profound stage of hypnosis, we may even be able to withstand the dentist's drill without flinching.

Many healers induce this theta state within themselves and their clients, promoting a sense of clarity and peace in the relationship. In the ideal encounter, the healer and healee, in equal partnership, gain insights that allow them to see beyond the confusion and chaos that confronts them. However, exactly how these messages are interpreted is of vital importance, requiring the healer to remain both rational and humble.

Recently, Alice, a woman in her sixties, came to me for advice. The previous year she had had a cancerous tumor removed surgically from the part of the bowel close to her anus. This left her with a colostomy, with her bowel opening to a bag on the surface of her abdomen. The surgeon was unable to remove all the tumor, and felt she was at risk of a serious blockage, or obstruction, if the opening of her bowel was not rerouted away from the anus. Naturally, the woman was upset by the prospect of having to have a colostomy for the rest of her life, and visited an internationally renowned healer for her advice. The healer, in a "theta state," informed her that the tumor had gone, and when shown a recent scan showing tumor, informed Alice that it was showing only dead tumor tissue. She encouraged her to request that her doctors consider a reversal of her colostomy, as she was now healed. The healer informed her that she had a 100 percent success rate. A follow-up scan showed the persistence of tumor, and Alice was keen to discuss the confusing situation she found herself in. On one hand, the healer was adamant she was healed; on the other hand, her surgeon felt she was at serious risk of suffering a bowel obstruction and breakdown.

My first comment was that, although inducing a theta state within us does often open our minds to valuable insights, even the most gifted healers, mediums, or clairvoyants do not achieve absolute success. Gifted remote viewers are accurate about a third to half the time, which is truly remarkable, and statistically very significant. However, would we trust a doctor if every other diagnosis he made was wrong?

And so I referred Alice to a senior surgeon, in my view a wise and humble man, for a second opinion. It is important that a patient in Alice's position is not ridiculed or criticized for seeking the advice of a healer, and that we as doctors understand the limits of our own knowledge. Alice's own involvement in her healing, her ownership, continues to be the utmost importance, and a health professional with an overpowering or patronizing manner could have a negative impact on her healing.

The original surgeon who performed the operation has remained totally supportive of Alice's journey, including her request for a second opinion.

Finally, Alice decided against a reversal of her colostomy, but the tumor has not progressed significantly over a year. This story emphasizes to me the importance of balancing intuition with rationality, and expertise with heartfelt compassion and humility.

Figure 26 reminds us how our intuitive mind in harmony with our heart allows us to be in a peaceful, receptive sate for healing. Figure 27 illustrates how we employ our rational mind to interpret our feelings, which present as subtle energetic variations sensed by our heart and brain. In this way, we relate rather than react to our feelings. This allows us to act positively and wisely within our time-space dimension, which feeds back into the field (or source) of consciousness, which is enhanced as a result. This model gives meaning to our actions, and offers one solution to the riddle "Why on earth are we here in this physical form?" We are here, it appears, to connect and act; this is the philosophical framework within which I am basing the chapters that follow.

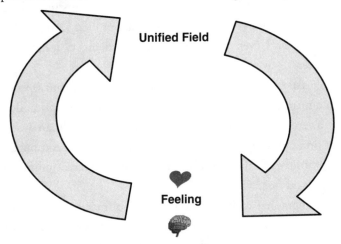

Figure 26. The intuitive mind—EEG in alpha or theta rhythm. Complete harmony "in the zone."

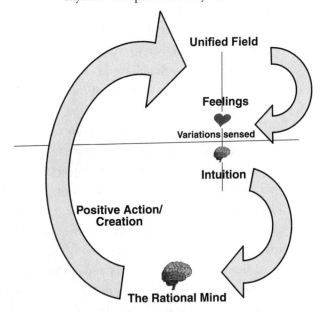

Figure 27. The rational mind—EEG in beta rhythm—responding to the intuitive heart/mind, which senses variations in the field. Rational thoughts lead to positive acts creating change in our space-time, enhancing the field of consciousness.

We continue to struggle to understand our most complex organ. No one can fail to be humbled by how little we do know of this amazing piece of hardware, which processes so much information in such a synchronized and efficient manner. It coordinates our most basic instinctual needs, the involuntary and voluntary processes within our body, while allowing us to think, plan, and even create. It allows us to move, feel, respond, and communicate.

It makes sense of the vague information presenting to our senses, focusing it all into a form that is useful for us. The holographic brain model suggests that this awareness, or consciousness, is due to more than just standard computer programming. Instead, our brains are quantum computers open to a universal library of infinite probabilities, exposing us to unlimited degrees of creativity, and free will.

The holographic model acknowledges the fractal reality of our brains. We humans are the most complex, chaotic of creatures on this earth. We are like the random twists and turns on the outer margins of a Mandelbrot set. The closer we look at our bodies, and our most perplexing organ, the more twists and turns we encounter. Our brain contains one hundred billion cells. Under a microscope, each cell appears as a beautiful fractal, each one instantly recognizable but completely unique. When looking at the fine tapered fingers of a dendrite, it becomes clear how something as small and compact as a human brain can gain access to such vast realms of information. And how each of us, as fractal beings, can make a difference.

And yet it is by releasing, temporarily at least, our need to know and understand, that we seem best able to access this wisdom. We must at times, quiet the chatter in our heads, and allow something simpler to unfold. In this way, we learn to worry less. We become less critical of ourselves and others, and less cynical about the ways of the world.

Those who have experienced an escape from their bodies (and brains) during a near-death experience (NDE) or out-of-body experience (OBE) frequently feel released from the burden of worry and self-criticism. Perhaps the most difficult part of the human hologram model to grasp is, not that the world around us is virtual, but that our own flesh, blood, and bones are somehow an illusion too. Or at least, only one version of a much greater reality. Someone who has separated from their body in an NDE or an OBE naturally grasps this concept more than others; no amount of study can compete with the wisdom gained from such an all-embracing experience. Often the experience is as real as any waking conscious event, and serves to put the troubles of the day firmly in their place. It also serves to reduce undue fear about dying, suggesting that something of us persists even after, as Hamlet put it, "we have shuffled off this mortal coil."

The theory of a holographic brain within a holographic human will remain controversial for some time. The concept that the world we perceive is a mere reflection of other truths challenges mainstream thought. For instance, if our perceptions of this world are indeed limited and self-selected, then it follows that realities must exist outside our current understanding.

This leads us open to the possibility that other such realities can be experienced while in an altered state of consciousness, for example, a shamanic trance. And that these experiences may carry truths, rather than delusions. Or that other intelligences more advanced than our own, or from our future, may be able to manifest themselves to some by breaking through the barriers of space-time. The possibility of hidden dimensions around us provokes intense interest in some, and intense fear in others.

The concept that our brains are complex quantum computers will frustrate those who are keen to simulate human intelligence and consciousness artificially in the form of machines inside and outside the body. But those who will continue to struggle the most will be those who feel that only their brains, and their brains alone, can possibly solve the mystery of their brains. For them the link between the open heart and the open mind may continue to prove a bridge too far.

Albert Einstein's brain was removed and preserved within seven hours of his death in 1955. Its architecture has been the subject of much study over the years, but apart from the recording of some unusual anatomical features, no great insights into his genius have emerged as a result. Perhaps the legacy of Einstein's genius lies not within a murky pot of formaldehyde on some dusty university laboratory shelf, but in his astounding life's work, and in the words of wisdom he leaves behind: "Pure logical thinking cannot yield us any knowledge of the empirical world. All knowledge of reality starts from experience and ends in it."

Chapter 16: Feelings

First feelings are always the most natural.
—King Louis XIV of France (1638–1715)

"Use your feelings, Obi-Wan, and find him you will."
—Yoda, Jedi Master, in *Star Wars Episode III: Revenge of the Sith*

First we feel. We are conceived on a feeling. In the womb, we feel soothed by our mother's heartbeat, but agitated by the sound of voices raised in anger. At birth, we feel the sudden shock of the unfamiliar and cry. As tiny babies, we cry when we are hungry, tired, and, as every parent will agree, for reasons unknown. Then we smile, and learn to share our feeling of joy by laughing. As toddlers, we feel frustrated and bored, and learn how to let everyone know about it. We laugh and shriek too.

As fully nurtured young children, we learn it is safe to feel and to express our feelings. By the age of ten, we are learning to relate, rather than react, to our feelings and to the feelings of others. First to those of our family, and then to those of our friends. This requires something new: reflective thought. So we may become shy, less expressive of our own feelings, as we become more "rational," some would say more sensible. As teenagers, we learn to feel love and rejection, to debate, and to disagree. As we mature, we grow intellectually. Our personality and ego grows too, securing within us a sense of self.

With balanced head and heart, we are then equipped to feel we know our special place in the world, strong in our resolve yet empathetic to the needs of others. We may now meet a soul mate, maybe a partner for life, together welcoming the opportunities and embracing the adventure each new day brings. As true renaissance women and men, we are now able to follow our true path in life, heeding Giotto di Bondone's call from over six hundred years ago to: "Take pleasure in your dreams; relish your principles and drape your purest feelings on the heart of a precious lover."

Well, in the perfect world this is how it would all unfold. For most of us, however, our early lives are more chaotic. And we can spend much of our lives lost within this chaos, struggling to come to terms with feelings hurt in the years before we learned to express them to others, or to rationalize them to ourselves. Maybe we were surrounded by others who themselves were overwhelmed by their chaotic feelings, unable to hear the cries of others. Only last week, a woman in her sixties with thirty years of extreme fatigue and widespread chronic pain explained to me how her father, a British World War II pilot, had been killed in a bombing raid over Germany, leaving her then heavily pregnant birth mother no option but to adopt her out when she was only a few days old. Despite kind and wonderful adopted parents, she could never remember a day when she hadn't felt lost and abandoned, feelings that only compounded when her dearly loved adopted parents died.

Sexual abuse shatters the sacred trusting innocence of a young child, infusing her or him with the abuser's pernicious intent, which, if left unaddressed, can lead to disharmony and disease throughout life. The invasion of a defenceless child's space often occurs at the most sensitive time in her life, a time before feelings can be easily expressed or rationalized. The scars, we know, last for generations, especially as abused children who have not been identified and helped have often turned into abusing adolescents and adults. Every month, I see adults, often with chronic ill health, bravely protecting their own children from a similar fate, effectively putting an end to a cycle of destructive behavior that has lasted centuries.

So it could be argued that feelings are more fundamental to our being than thoughts. From the basic survival instincts of fear, pain, hunger, and thirst that we share with much of the animal kingdom, to the feelings of compassion and empathy for all life forms that must now form an essential human trait if life on Earth is to continue to thrive. It is only once we are comfortably secure in our feelings that we can take the positive steps needed to ensure that this happens. We may need to express these feelings emotionally and rationally to others, sharing our vision, inspiring others to work in collaborative teams. Ultimately, however, actions speak louder than words. It was Florence Nightingale, the pioneering nurse whose selfless work saved countless lives from injury and disease in the Crimean War, who said: "I think one's feelings waste themselves in words; they ought all to be distilled into actions which bring results." She famously earned the title "The Lady with the Lamp," as she would continue to make her rounds, alone and well into the night, long after the medical staff had retired to bed.

Yet, despite the obvious vital importance of feelings to the human condition, little can be found on the subject in modern medical and psychology textbooks. Despite the fact that for thirty-five years, every single person who has presented to me in my medical practice has first and foremost told me how they feel. And I suspect I have greeted the vast majority with the words: "How are you feeling today?" The answers: "I feel tired/in pain/sad/worried/desperate" and occasionally (thankfully), "a little bit better."

At least now we are beginning to recognize there is a problem. The Australian philosopher Professor David Chalmers in his 1996 book *The Conscious Mind* identified that there are both "easy" and "hard" problems to be solved if we are to understand the true nature of human consciousness. The "easy" problems are those accessible to objective scientific inquiry, such as how information is processed by the brain. The "hard" problems are all about subjective human experience. For example: Why does a high C from Luciano Pavarotti send shivers down one's spine? Why does the deep orange of a sunset enrapture us so?

It would seem that these "hard" problems are all concerned with our feelings, and that, ironically, a child, a shaman in the Peruvian rain forest, or even possibly a chimpanzee would find accepting such subjective experience neither "hard" nor a "problem." For them, the problems identified by the philosophers as "easy" would probably prove far more difficult to grasp. In the modern world, we are conditioned through our education to apply our logical, rational minds to solve such problems. If feelings and experience are more fundamental than logical thought, then maybe (and logically!) we will most likely find the answers by first examining our own feelings and subjective experiences. Thereby our cognition, our thinking, can resume its rightful place—deciphering the messages we are receiving in a pure, but as yet, immeasurable form.

The human hologram model acknowledges that information plays a primary, fundamental role in our universe, and that another description for this universal information is a *unifying field of consciousness*. We receive this information in its purest form through our feelings, our instincts, and when our heart is open, our intuition. When fully relaxed, in the zone, away from our worrying active mind, we can wallow in its timeless peace. We can remind ourselves of its presence, and recharge our being. Ultimately, however, it is unlikely we are here on this planet to exist perpetually in this state of bliss. We are here to act and contribute.

I theorize that our feelings represent subtle fluctuations in the field of consciousness. We can now speculate on the mechanism by which our heart receives and transmits such fluctuations, relaying them through our body as messages we identify as our feelings. To understand this, we must entertain the notion that this, our most electromagnetic organ, can achieve the same perfect balance that is known to occur between two entangled atoms. In the simplest case, for example, between two entangled hydrogen atoms, the spin of the electron in atom (a) is perfectly balanced by the opposing spin in atom (b). (See figure 28.)

Note: Electron on left spins clockwise; electron on right spins anti-clockwise.

Figure 28. Entanglement between two electrons in a hydrogen atom.

It is only a small step, figuratively, from this simple line diagram of entanglement to the familiar symbol of infinity. The similarity of these two forms appears to be no coincidence. The energies—positive and negative, yang and yin—can be perceived to balance perfectly at the central crossing point. (See figure 29a.)

As all electrical forces are balanced out at this point, potentially we have a vacuum or portal to nonlocal information—in other words, instant holographic coexistence. This proposed phenomenon has attracted several labels including the "zero point of the heart" and the Möbius heart.[84] We can create a version of a Möbius strip by forming a figure eight from a piece of ribbon, making sure we twist the ribbon through 180 degrees before

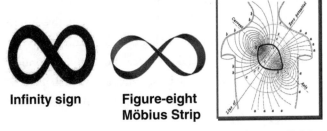

Infinity sign   **Figure-eight Möbius Strip**   **Magnetic field of the heart**

Figure 29a, infinity sign; b, figure-eight Mobius strip; and c, magnetic field of the heart (Augustus Waller, 1887).

connecting the ends. The other name for the infinity symbol is the *lemniscate,* from the Latin *lemniscus,* meaning "ribbon." It is worth noting, however, that in popular culture the infinity symbol preceded the Möbius strip by several centuries. (See figure 29b.)

Theories abound about just why the geometry and anatomy of the heart makes it the ideal structure to receive and process nonlocal fields. Most poignantly, the strong magnetic field of the heart, like that of our earth, has opposing poles, with a central "zero point" plane of perfect balance. (See figure 29c.) But there are theories too that may add additional insights. In Vedic thought, the heart chakra is found in the central position—chakra four out of seven. Sitting in the lotus position, the body adopts the shape of a pyramid, with the heart in a central position. The word "pyramid" means "fire in the middle." The whorled spiraling of heart muscle, the interacting currents of blood flow within the chambers of the heart and great vessels, the geometry of the blood vessels, and the balance between oxygenated and deoxygenated blood (and iron) are all factors that some speculate are important. For all these reasons, I have depicted, in the diagrams throughout this book, the heart as being the focused receiver of information from the unified field.

Once received, the nonlocal information is "then" relayed by the heart both locally and holographically to the rest of the body. Local transfer occurs through chemicals and via the waves of pulsed energy, which spread through the body and interact with the energy fields throughout the body. There is thus a sharing of information, with interference patterns forming, as described earlier. We now have a coherent environment within the body as a whole (through the canceling out of forces known as destructive interference) for further "entanglement" with other fields of information, and their ultimate source; in other words, our body's nonlocal holographic connection to the unified field. Hence, in theory, it is our heart that first receives, and then relays to our body and brain, subtle changes in this field, which we each record as "feelings."

There is even experimental evidence that supports *precognition,* where the heart and brain can sense an event prior to it being perceived as happening. Moreover, this research suggests that the heart receives this intuitive information before the brain.[85] This holographic model also acknowledges that information fields emanate from our hearts and bodies into the environment, again both locally and nonlocally. This, in theory at least, lends some credence to the concept of prayer and global consciousness.

So it makes some sense that it is our hearts, our most electrically powerful and sensitive organ, that process this information, transforming it into feelings within us. In their simplest, most instinctual forms—those of fear, hunger, and thirst—our feelings are messages that prove vital to our survival. We can react to these appropriately, some would say subconsciously, without the need for our advanced powers of cognition. However, other feelings, of love, of vocation, and of higher intuition benefit us more if, rather than simply react, we take time to relate to them. In other words, we learn to use our heads to listen to our hearts.

The human hologram model honors our feelings and our thoughts equally. For it is by thinking things through that the appropriate actions can be taken to achieve the balance that is needed within the field of consciousness. We are thus responding appropriately to the messages we are receiving through the energy of our actions.

So it seems as we struggle from day to day within our unique four-dimensional lives of chaos, we are really highly sophisticated bioconscious feedback machines, helping to refine and renew the very source from which we emanate. We need the chaos, the challenges, the debate, and the dissent to help focus us at this, arguably, the most critical time of our evolution.

There are those who know these truths, and they are our teachers. Mark Twain once wrote that "really great people make you feel that you, too, can become great." I can think of no more inspiring words to end this chapter on feelings than those of Mahatma Gandhi, whose life was dedicated, and ultimately sacrificed, to the achievement of peace through non-violent protest:

"I offer you peace. I offer you love. I offer you friendship. I see your beauty. I hear your need. I feel your feelings. My wisdom flows from the Highest Source. I salute that source in you. Let us work together for unity and love."

Chapter 17:
The Human Hologram and Healing

*Observe, record, tabulate, communicate. Use your five senses
...Learn to see, learn to hear, learn to feel, learn to smell, and
know that by practice alone you can become an expert.*
—Sir William Osler (1849–1919), Canadian physician

We call them symptoms. These are the confusing, random bits of information coming from their bodies that patients try to describe to their doctors, people who are trained to make some sense of them. Symptoms are feelings—subjective and rarely measurable objectively. The only real evidence that symptoms, or happenings, such as pain exist at all is that we all experience them.

Sure, we now know through sophisticated computerized imaging techniques that a certain part of the brain lights up when we feel pain in a certain part of the body. But we need to remember that this image isn't our pain—simply an indication that our brain is somehow involved in the process. However, one such imaging technique known as functional MRI (fMRI) is confirming that the more sensitive amongst us can feel the pain of others. In a 2009 paper published in the prestigious journal Pain, researchers in Birmingham, England, showed a sample of college students images of people receiving injections and suffering from specific sports injuries.[86] Nearly a third of the students could actually feel the pain in themselves, at precisely the same site portrayed in the photograph they were viewing. The researchers then performed fMRI scans on the ten students who felt the pain, and compared them to ten of those who had only reported an emotional reaction to the images. The results were interesting. Whereas all students showed increased blood flow in the areas of their brains known to process emotions, only the students who could actually feel the pain had activity showing up in the specific pain-related areas of their brains.

The researchers are now interested in discovering whether those who suffer from the chronic pain or fibromyalgia are more likely to be those who actively feel the pain of others. My own observation over the years as a doctor, and as a sufferer of fibromyalgia myself, suggests that this is highly likely to be the case.

This research provides hard evidence that pain is not confined within the skin of one person. And although it is doubtful that we have literally to feel the pain of others to be truly empathetic, it does show that feelings are universally shared. Other researchers have suggested that feeling the pain of others "by proxy" is an essential survival mechanism. In a recent experiment that would alarm animal rights activists, researchers studied the brain activity of one mouse watching another being given electric shocks.[87] The observer mouse's brain showed activity in the specific region (the amygdyla) known to be associated with fear and with learning how to take the appropriate action in future fearful situations. The other region showing activity was the anterior cingulated cortex (ACC), an area known to be associated with the emotional aspects of pain. It is speculated that the observer mouse computes and remembers his unfortunate friend's suffering as if it were his own, allowing him to take evasive action when potential conflicts are brewing. (I would suspect that by now he would already be plotting his escape from the laboratory concerned!)

So for mice, at least, empathy for the plight of fellow mice may play a valuable role in their survival. In the following pages, I will suggest that we can now use our deeper understanding of feelings to help others heal. And I'll discuss how an empathetic, compassionate approach to medicine, healing, and life is, beyond doubt, the way forward.

Over the past two decades, the pace of our lives has accelerated. We have more to learn, more facts at our fingertips, more choices, and more confusion. We have more to worry about. People presenting to me now feel not only sick, tired, and in pain, but also worried, confused, and frustrated. Why can't someone tell me what's wrong? Why do I have to take all these pills? What if I lose my job? Their worry confusion and frustration is impacting on their already delicate immune systems, making them even more sick and tired and worsening their pain. Trapped in a compounding cycle of misery, they can't sleep, they feel depressed. In short, they are thoroughly sick and tired of being sick and tired.

But before you send me cards of sympathy for my role in all this despair and suffering, let me reassure you that deep within each of these unfortunate visitors is a spark of hope. And there are yet others who, despite their considerable ailments, inspire me beyond measure by their uplifting take on life. One in particular, Dianne, who despite witnessing the relentless spread of melanoma into every corner of her body, was able to make me laugh (and cry) every time I visited. Her immense sadness about leaving her family behind was balanced by an intense curiosity (an exhilaration even) about what she would experience in the process of dying. A process she viewed as a mere transition of form. And, as she would tell me, in the unlikely event this proved not to be the case, and her essence completely ceased to exist, well, there would be nothing left of her to worry, would there?

So let's start with a simple model of healing, something we can apply to those whose conditions have not yet compounded through worry and confusion. Let's start by examining what happens when we stroke a puppy, adapting the feedback model introduced on page 21. (See figure 30.)

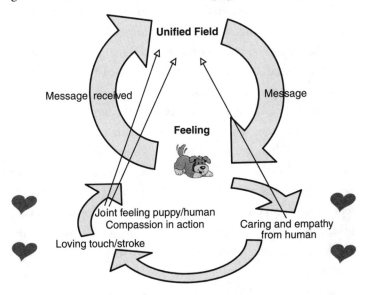

Figure 30. Stroking the puppy.

This is a spontaneous heartfelt act that requires little thought. Yet it brings great satisfaction to both ourselves and the puppy. Only good can come of it (definitely a positive contribution to the unified field), from that impulsive moment when we decide to stroke to the stroke itself—a gentle act of compassionate connection.

There are ongoing spin-offs from this simple act. If other puppies in the litter are watching, they will want to be stroked too. Together they learn to trust, not to fear, humans. And as for us? Well, we just love stroking puppies. And there is now good evidence that dog owners lead healthier lives, with lower blood pressure readings, lower cholesterol levels, and fewer symptoms of stress.[88]

Of course, even more profound than the special relationship we nurture between young animals and us is the special bond we form with the young of our own species. A mother responds instinctively, and intuitively, to the feelings of her baby. A hungry crying baby is soothed on the breast of her mother, who, in turn, feels deeply and lovingly connected to her baby. As babies, we were introduced to this pure loving connection, and using our holographic model, the source of our being, soon after we arrived on this earth. For all of us in early infancy, it was our heartfelt feelings that mattered, feelings that we communicated directly to our parents who responded instantly with maybe a hug, maybe a drink, or maybe even a fresh dry diaper. These were the times when our connection to a benign field of consciousness, first planted as a seed at conception then nurtured in our mother's womb, was kept alive as we prepared to take our first steps on our planet. As parents, teachers, healers, and responsible adults, our primary role is to ensure that these loving connections remain intact, and strong enough to last a lifetime.

As young children, we were open and vulnerable. Our rational focused mind, as depicted on an EEG as a beta wave pattern, was yet to develop. We were carefree, creative, open to flights of fancy and fantasy. We could feel deeply, and show our feelings, but not yet add logic to make sense of our world. If something happened, for example, a car crash or a house fire, we likely felt it was our fault, unless supported by a caring patient adult. Our brain waves were predominantly of the alpha or theta shape, so we were highly suggestible, and hypnotizable. Those children exposed to frequent episodes of fear—for example, constantly having to witness the aggressive behavior of intoxicated parents—can absorb this fear, and be conditioned to react in this way throughout their lives. Hence the patterns of addiction and abuse continue to be seeded and reseeded in generation upon generation.

On hearing patients recount such stories, and on realizing that such cycles of toxicity have been present in their families for centuries, I am frequently reminded of F. Scott Fitzgerald's final sentence in his celebrated classic novel *The Great Gatsby:* "And so we beat on, boats against the current, borne back ceaselessly into the past." Once these cycles are recognized, however, they can be terminated for good. I have witnessed many occasions in which a mother, abused as a child, creates an environment for such deep healing to occur. On occasions, the mother has had to escape from a destructive relationship with a partner, himself the product of generations of abuse. And yet, through the mother's courage, compassion, hard work, and intelligence, the cycle comes to an end as she protects and nurtures her children anew. Yet often it is her own health that she neglects, and hence the healing is only complete when she learns to show compassion to herself. Sometimes this lesson is learned as she herself confronts an illness, and realizes she must receive love and compassion too. These are teachings she missed out on when she was a child, as her voice strained to be heard.

Ruth, a woman in her late eighties, consulted me recently. As is my usual practice, we discussed her childhood years in detail. She told me that her father was a womanizing, verbally aggressive drunkard, and that from the ages of three till eight, the time her parents separated, she was completely mute— in retrospect, a natural defence mechanism. Naturally, she had found great difficulty throughout her adult life in forgiving her father for his indiscretions. Until, that is, she met for the first time her father's sister while visiting her relatives in Liverpool, England, many years later. Her aunt recounted that when her father was fourteen, his own father, as punishment, had placed him on a merchant ship bound for Sydney, Australia, with two pounds in his pocket and no arrangements made for accommodation at the end of his journey. This was in 1910, and it appears several of his fellow travelers did not survive the long journey. On disembarking in Sydney, with nowhere to go, he found a park bench and fell asleep. He was awoken by a kindly tramp who, on learning his fate, notified a social worker from the city mission.

As a result, he was clothed and fed, but sent to the outback of Australia to earn his keep on a cattle station. Surrounded by hard living cattle-hands— he was the youngest by ten years—he learned to work hard, to fight, and, inevitably, to drink.

On hearing this story from her aunt, Ruth was able to understand her father's plight and to achieve a level of forgiveness for his actions. Throughout her life, she had been plagued by throat problems, with a difficulty in swallowing, that had required many investigations and medications. These symptoms had improved somewhat as she learned to forgive her father, but were still with her. We now hope that, by allowing her to express herself, even eighty years after her trauma, there will be further healing.

It is clear every child needs her parents to be a supporting, affirming presence in her life. As our child grows, we learn to listen to her. Where does it hurt? Where did you fall over? Our child needs to be heard, to understand that expressing her feelings is important, as she learns there are great benefits in sharing information with others. The fuss, the cuddles, the Band-Aid—all form part of the ritual that eases our child into the rough and tumble of life on Earth. (See figure 31.)

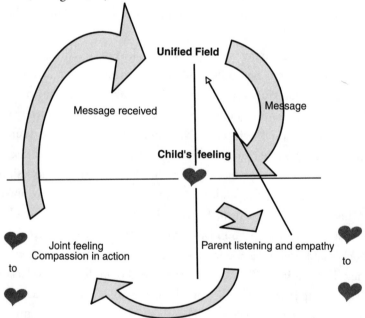

Figure 31. Parent–young child healing bond healing young child.

If this does not soothe our child, then as parents we have to use our brains—maybe we need this checked out, possibly see a doctor. The child's feeling then becomes a "symptom" (or happening) to the doctor, who may organize an X ray or apply a special dressing to a deep wound. Often, by the time the parents have traveled to the clinic, filled in the forms at reception, and waited ten minutes, everything has settled. The doctor is then there to add the final reassurance. If however, the pain persists and the child is unwell, then the doctor's training becomes important—the X ray, the bandage, and the cup of tea for the parents. The cycle is complete, with rational thought and medical training being used appropriately, thus leading to peace and healing. (See figure 32.)

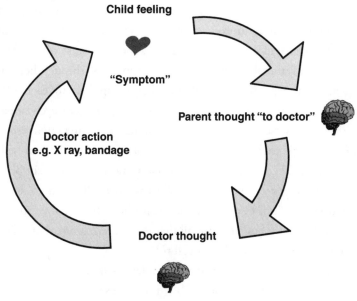

Figure 32. The rational medical mind.

I have introduced this simple everyday medical scenario as a starting point in explaining how the human hologram model relates to my own life. Over the following pages, I hope to show that an awareness of the hidden dimensions explained so far in this book could well have real practical benefits for us all. I can only talk with any authority about my own experiences; I warmly invite you to apply any new insights gained to enhance your own life, and the lives of your loved ones. The patterns and cycles presented throughout the book are universal, fractal, and holographically related. They represent an underlying symmetry and balance often concealed from the chaos of our everyday lives.

As medical science progresses and our knowledge expands, we are witnessing doctors having to dedicate their lives to ever new sub-branches of specialties. For instance, whereas thirty years ago there were heart specialists who were all experts in diagnosing heart conditions, prescribing heart medications, and reading cardiographs, now there are heart specialists who have to devote their whole lives to interpreting the images seen on complex and sophisticated computerized scans. We need these highly skilled, clever, dedicated people to make sense of the chaos we are uncovering. In all industries, we are seeing this specialization—expertise in diversity, each job as important as the next in expanding our knowledge, and contributing to a more joyful, peaceful existence here. Each super-specialist who follows her true vocation and heart is expanding our consciousness through her actions.

However, the process of healing, whether applied to each of us as individuals or as part of a global family, shows a consistent cyclical pattern. This pattern reveals itself as we grow up and mature into integrated responsible adults. In my work, each time I am meeting someone anew, I ask myself what role I am being asked to play in this person's healing. Am I simply being invited to use whatever expertise I have learned over the years to decide the correct course of action? For instance, is an antibiotic needed for a sore throat; does this cut need stitches? If this is all that is required of me, then this is fine and carries with it much satisfaction all round.

On another level, however, if the problem is more complex, I need to know the answers to more questions besides. For example, am I being asked to take complete charge, like the parent of a small child who is continuing to cry, or am I being asked to form an adult partnership for joint discussion and action? Am I being asked for a rescue or for a reassurance? What part, if any, will my presence play in the person's healing? And is this what is wanted?

If, indeed, a rescue is needed, it is an indication that the person in front of me, initially at least, is coming as a small child; trusting, passive, yet vulnerable. Compassionate listening is in order, but with the awareness that ideally this "child" may need help in growing up, in taking full responsibility for his life. This may initially prove to be a rocky road we walk, as the "child" travels through an awkward adolescence.

The Healer as Parent

It is perhaps obvious that any deep healing process involves revisiting our past, retracing our steps from early childhood to adulthood. So often we need to heal injustices suffered as children—feelings of abandonment, unresolved grief, and the disempowerment and emotional scarring associated with physical and sexual abuse. There are many other, perhaps less obvious traumas that need to be revisited and healed. The firstborn child who has to sacrifice a carefree childhood of playfulness to adopt a role suited to someone far older. The gifted child driven by ambitious parents to achieve academically at the expense of the spontaneity and wonder of childhood. The child who is not permitted to experience failure and the child who hears constantly that she will always fail and "never amount to much."

We all have a responsibility as fellow adults to help each other heal from these indiscretions in our early lives. All of us go through life revisiting these issues; some of us are lucky enough for them not to dominate our life and health. Those in the healing and teaching professions have particular responsibilities, as adults involved in overseeing someone's physical and emotional growth. In many cases, I have come to regard my role as a doctor as a responsible parent, with all the privileges and problems that accompany that very special role.

Significant personal growth always accompanies deep healing, and so it is perhaps inevitable that the person helping alongside adopts a parenting role, a role that may have to change and adapt very quickly as the " child" evolves rapidly.

The role of the healer, for those who first and foremost need a rescue, those who utter a cry for help, mirrors the role of a parent to a young child (see figure 31). Trust needs to be engendered, and true empathy shown. There may be an early discomfort about receiving such attention, as the person may have such low self-esteem that they do not feel worthy of such attention. However, it is important that the next step—their own involvement and say in their healing—is facilitated as soon as the person is ready, so as to avoid a state of dependency on the healer. A significant part of growing into balanced adulthood involves learning to take full responsibility for our lives. In practice, this means working through our feelings, expressing them appropriately through our emotions,* and then applying our rationality, our thoughts, so that we can respond effectively, rather than simply react to whatever life serves up. Figure 33 illustrates the next step.

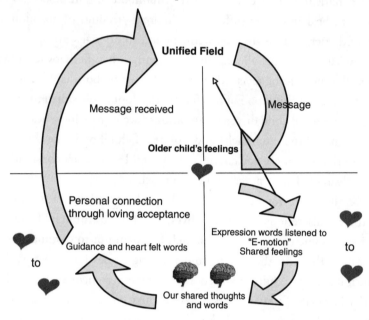

Figure 33. Parent–older child healing bond.

This ideally is how a parent interacts with an older child or teenager. A joint solution is sought, nurturing within the child an inner confidence yet maintaining an openness to ask for help. Of course, as with bringing up teenagers, the healer can expect disagreement, manipulation, and inconsistency along the way! There have been occasions over the years when a patient has made a far from graceful exit from my office, reacting to my suggestion of a true healing partnership. But like an errant teenager who storms out of the house, slamming the door behind him, he returns when the time is right. (With the male teenager, this inevitably coincides with the very first pangs of hunger!)

In my medical practice, I find acupuncture very useful in this intermediate phase of healing. The person is becoming more involved in his healing,

---

*The word "emotion" is derived from the French verb esmovoir translated as "to set in motion" or "move the feelings."

as he has presented for a treatment that relies mainly on his own inner potential to heal. He learns to feel at peace and relax, albeit with my help. This is also a time when he can express his feelings, if need be, in a safe, trusting environment.

With all these factors in place, it is highly likely that underlying fluctuations in the field of consciousness that manifest as feelings or symptoms are acknowledged and answered along the way. This feedback process (shown in figure 34) results from a joint harmony of hearts and heads.

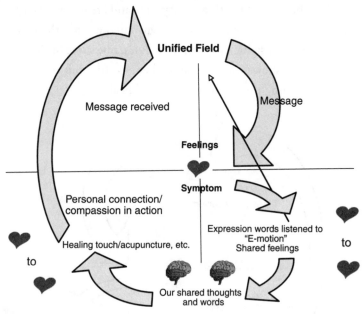

Figure 34. Healer as parent to adult—stage 1.

Thus the perfect environment is created for the next step to be taken, when the time is right. This is the step from a healing adolescence into a healing adulthood—full responsibility—which, of course, still includes asking for help whenever this is needed.

How we behave in the presence of an adult who has suffered from abuse or belittlement as a child is of primary importance. The more that compassion and tolerance is experienced firsthand by the person, the more confident and the less vulnerable that person becomes. Simple breathing and meditation techniques can be taught early on in the process. This has the effect of reducing the number of visits needed, as self-healing largely replaces healing from another party. Ideally, the person learns, simply from being in the presence of the healing practitioner, how to become compassionate and nonjudgmental towards himself. Energy psychology exercises, including the popular Emotional Freedom Techniques (EFT), are an excellent way to help someone take control in this way. In my own practice, EFT represents an ideal next step, as it relates directly to traditional acupuncture points and the Chinese meridian system. Those receiving acupuncture have already experienced the benefits of treating these points and are naturally enthusiastic about the prospect of influencing their own "energy grid" themselves, this time with no needle in sight. In appendix 1, I describe some of these exercises in detail. Figure 35 shows this next step taken, as the adult healee is guided through a deep healing process by the "parent" healer.

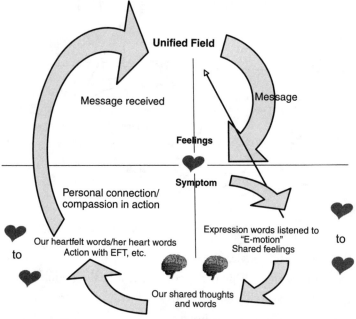

Figure 35. Healer as parent to adult—stage 2.

Of the utmost importance for adult healing is the following sequence: feelings acknowledged > emotions expressed > rational solutions sought > correct action taken. This order allows us to relate, rather than react impulsively to our feelings—a pause for thought about the effect our actions will have on others. If someone abuses us, then rather than react instantly by abusing them back, we find ways to express our displeasure and then rationalize how we will manage the situation in future. If we are guilty of beating a fellow driver to the last parking space outside the supermarket, a driver who we realize upon reflection has been waiting longer than us, then we should probably respond gracefully to his forcefully expressed displeasure by backing out and vacating the space. If however, we are convinced that he is mistaken, and that we through our patience who have the right to occupy the space, then we should maybe only offer back an expression, a friendly wave or maybe a shrug, that conveys our concern about his plight but puts the whole affair into perspective. On such an occasion, I would not recommend the tactic of blowing a kiss—a response I have found rewarding on the rare occasion I have been the recipient of the raised middle finger from an enraged fellow driver—as it is highly likely that we will meet face-to-face only minutes later in the aisle of the supermarket!

On a more serious level, though, disruption to this sequence may have a major impact on our health. It is important as we grow into adulthood that examples are set for us by our parents or guardians who themselves have learned to express their feelings, listen to others, and to take full responsibility for their actions.

This week I met, for the first time, Malcolm, a self-employed plumber in his late forties. He had a very rare form of cancer in the bones of his back and pelvis, and had for several months undergone several cycles of chemotherapy. Despite tissue samples of the tumor that had been sent to overseas' experts, there was still no definitive diagnosis. As the tumors were slow to respond to the chemotherapy, he was having to decide whether he would embark on more cycles, this time with even more toxic drugs. None of his doctors could tell him how long he had to live; they could only say that it seemed a particularly aggressive, poorly responding cancer.

He came to see me with his wife, Ella. They seemed extremely happy together and very supportive of each other. However, Malcolm opened the discussion by explaining that they had separated a year before the diagnosis, and for months he had felt lonely and depressed. Wisely, during this time, they agreed to extensive counseling with the result that they reunited, with many major issues resolved. It was soon after this that his rare bone cancer was diagnosed.

He also told me that at the age of twenty-four, after a painful kick in the groin by his karate instructor, he went to see his doctor who diagnosed a testicular cancer. This was treated and he was effectively cured; the new tumor that he had now was not linked to this disease.

At this point in the consultation, I explained that there was good evidence that stresses in middle life, especially feelings of abandonment, often relate back to traumas in childhood that lie buried, yet to be resolved. I explained further that such traumas could lead to genes expressing themselves as physical illness, and that, as we age, our immune systems for many reasons may not be able to cope as well as before.

I asked Malcolm about his childhood, in particular the times that he may have felt abandoned or lost. He told me that his parents had split up when he was ten. His father was a sales representative, and the family had never stayed in one home for more than two years. He remembers taking the news of his father's leaving (for another woman) very calmly, rationalizing that it would make little difference since he had seen so little of him anyway. He was the oldest of three children and remembers being thought of as sensible and mature for his age. There had been no abuse in the family, no fights between his parents. However, the situation was to change significantly with the arrival of his mother's new partner, a farmer. Initially, all was well, and Malcolm remembers loving farm life and the stability of living in one place and attending one school for several years.

During the first year, he noticed his mother was collecting bruises, which she explained away as due to falls and mishaps around the farm. The physical abuse at the hands of her new partner increased, and after two years she left home together with her two daughters. Malcolm stayed on the farm. Despite becoming aware of the trauma suffered by his mother, he had formed a close relationship with his stepfather, much closer than that he had experienced with his birth father. He had grown to love farm life and had many friends at the local school.

After only a few months, his birth father reappeared and forcefully took him from the farm, back to the city. According to both his birth mother and father, his stepfather—and there was some truth to this—was mentally ill, and this move was thought to be in his best interests. Malcolm learned to cope with city life again, and at the age of eighteen enrolled to learn the trade of plumbing. He met Ella, his wife-to-be, soon after this.

As Malcolm recounted his life story, I asked occasionally if he felt any of this was relevant to the situation in which he now found himself. On each occasion, he nodded. It would appear that deeper meanings were emerging, links that made some sense of his plight. We seemed to be addressing a recurring feeling of being lost and alone—a feeling further compounded by his present serious and mysterious illness.

We are still at our infancy of understanding how best to advise Malcolm in the situation he finds himself. I would suggest that the most important first step is his awareness of how his current crisis relates to his unique life story. In my experience, making these connections empowers the person by giving them a framework, often for the first time, in which to plan a program of deep healing. They take full adult ownership, and control, of their lives.

There will be critics who will point out that there is no objective proof that the traumas of Malcolm's early life, and even the crisis that he experienced the year before his diagnosis, have any relevance to his illness. That fanciful, irrelevant associations are being made that, at best, are a serious diversion from the more important role of aggressive drug treatment. There will be still more who ask: "So what if Malcolm had a tough time in the past, why waste time (and Malcolm's money) on something that can't be fixed? And anyway, what on earth can anyone do to help?" In reply, it is important to point out that there is now mounting scientific evidence that stress can play a major role in both the development and the progression of cancer.

In the past few years, it has been shown that:

1. Cancers of all types are more prevalent in holocaust survivors.[89]
2. A signaling process in cells known as JNK, recognized as being induced by emotional and physical stress, causes genes to mutate into cancer-forming genes.[90]
3. Increased levels of epinephrine, or adrenaline, induced by stress cause ovarian cancer cells to spread, or metastasize, away from their primary site. High levels also activate an enzyme, FAK, that stops the body from destroying ovarian cancer cells.[91] Yet another enzyme, BAD, that causes breast and prostate tumor cells to die is inactivated by high adrenaline levels.[92]

4.   Research shows that married cancer patients who are separated at the time of their diagnosis do not live as long as those who are widowed, divorced, or have never married.[93]

It is important to recognize that, alongside the emotional stress, other associated factors may contribute to a weakened immune system. For example, those who suffered the extreme emotional hardship of Nazi concentration camps were also denied adequate nutrition, medical care, and sanitation. Even in the less extreme scenario of a marital separation, the person's diet and self-care can suffer, laying the body open to disease. It is apparent, however, that in all these cases, it is a disruption of the feelings of the person that lies at the formative root of the process.* The chemical changes, such as an elevation in the body's adrenaline level, are simply the body's response to these feelings.

So from the accumulated data, it is becoming clear that both past and ongoing stresses should be addressed as soon as possible when someone presents with cancer. To help someone who finds himself in Malcolm's position, there is now a range of psychological services available that are not merely there to help him cope; they may also play a significant role in strengthening his immune system.

In Malcolm's case, the loneliness and stress of his separation likely played its part in the advent and spread of his tumor. As the manner in which we respond to such stresses is conditioned in us from an early age, then it makes sense to explore and heal the roots to the best of our ability. Energy psychology techniques such as EFT and TFT (Thought Field Therapy) have clear advantages as, once learned, they are simple to use and are completely free of charge—something much appreciated by those chronically ill, so often sapped of energy and finances. Although studies involving these techniques have yet to be performed on cancer patients, promising results are being obtained on U.S. veterans of the Iraq and Vietnam wars suffering from anxiety, depression, and posttraumatic stress disorder. A small pilot study published in 2009 showed significant improvement of symptoms after six sessions of EFT.[94]

---

*In all cases, underlying environmental toxins should also be sought.

And so I am encouraging Malcolm to perform simple exercises twice a day that address an emotional state that may have played an important part in the development of his condition. These involve crossing his hands over his heart while breathing gently into his abdomen, and saying certain heartfelt phrases. The phrases come in two parts. The first part relates back to his body, in the simplest terms possible, the difficult situation in which he finds himself:

"Even though I have this rare form of tumor…"

The second part is a simple expression of compassion to himself:

"…I deeply and completely accept myself."

In addition, Malcolm is to say, "Even though this followed a time I felt lonely and abandoned, I deeply and completely accept myself."

These simple exercises, used alongside supportive counseling sessions, help demystify the situation in which Malcolm finds himself and help resolve any deep feelings of frustration, confusion, and guilt.

After a week, we can introduce some other phrases that relate to the traumas experienced in childhood. In my experience, and that of other experienced therapists, this approach, if conducted with sensitivity, seldom evokes a negative abreaction as the potentially painful memories are uncovered. I do however, encourage gentle—almost poetic—phrasing, and we may initially avoid the word "cancer" if the person associates this with negative images and feelings of disempowerment. (For more details of these exercises, see appendix 1.)

Perhaps the most important step over the next few years is for everyone presenting with a complex illness to be given the opportunity to explain his or her life story to a suitably qualified professional, in a peaceful and unrushed environment. Then a suitable plan can be made to address and heal any unresolved issues, alongside any medical or surgical care that may be thought appropriate.

But it is the ultimate responsibility of each one of us to ensure that unresolved issues and harmful negative beliefs are healed within ourselves, our children, our friends, and all those seeking our help. Only then, will the toxic and addictive cycles of behavior that have plagued humanity for millennia be broken for good.

Chapter 18:
The Human Hologram
and the Medical Model

It was the Father of Medicine, Hippocrates of Kos, who offered up the following sound advice some twenty-four centuries ago: "Prayer indeed is good, but while calling on the gods a man should himself lend a hand."

If today I suddenly were to begin to suffer from excruciating toothache, I would not head forthwith to the nearest peaceful glade alongside a babbling brook and seek coherence with the universe. I would make an urgent appointment with an emergency dentist. I would care less whether she or he shared my holistic philosophies, and far more that she or he was suitably qualified and was, naturally, available. It would, of course, be ideal if she or he didn't have an anger management issue, as in my experience, angry dentists, together with angry acupuncturists, are best avoided. But even this preference would take second place to making sure, to quote Macbeth, that "If it were done when 'tis done, then 'twere well it were done quickly."

With my tooth drilled and filled, it is likely I could return that day to my work, or possibly even to the golf course. The job performed by the dentist would then be deemed a benevolent act, and whether the dentist was aware of it or not, would enhance the unified field (as would my paying his fee with speed and grace).

Similarly, doctors trained primarily to react with skill to emergencies are tremendously generous contributors to the field, as Florence Nightingale would surely have agreed. Focused thoughts leading to focused actions are at play in these jobs, and many years of study, dedication, and sacrifice lie behind the efficient running of every ER department.

As we learn more and more about the complex mechanics of our bodies, medical specialties have further divided into subspecialties. The techniques and equipment used to ensure that investigations are accurate and comfortable are now so sophisticated that they are only truly effective, and safe, in the hands of those using them on a regular basis. These super-specialists are each dealing with their own tiny but vital branch of the ever-expanding complex whole. I envisage their work as existing metaphorically at the fringes of a Mandelbrot set, the ever-changing complexity we find as we delve deeper and deeper into the chaotic peripheries of our world.

Again, if I was unfortunate enough to develop chest pain, and one of my coronary arteries was shown on an angiogram to be blocked, I would be the forever-grateful recipient of a tiny stent that can now be put in place via a flexible tube passing through my blood vessels. Patients suffering from the constant pain of arthritis in a hip can gain instant relief as soon as this joint is replaced. Acupuncture, although offering significant temporary relief for those suffering from this pain, cannot compete with surgery performed when the time is right. As indicated in figure 36, in all these cases, focused rational thought patterns in the brains of the doctors and surgeons, as depicted on an EEG as beta wave activity, are essential. A steady hand is also of prime importance.

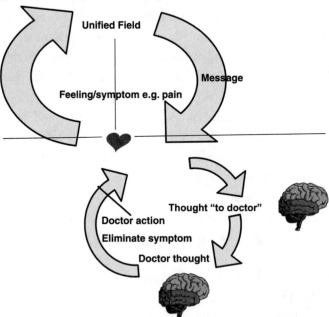

Figure 36. The medical model.

In each of these cases, the person undergoing the rescue procedure has submitted to someone who has the necessary training, skills, and intelligence to fix the problem. Ideally, the person does this consciously, fully informed of the risks and potential benefits of the procedure. Once alleviated of the symptoms, the person can feel well, engage in life more fully, and, therefore, as illustrated in figure 36, potentially be in harmony again with the unified field. However, many of these mechanical problems have presented merely as the end-term consequence of lifestyle and relationship imbalances, often with their roots, as we have already explored, in someone's feelings and cravings.

The Western medical model, for the past sixty years at least, has focused on how best to rearrange the chemistry and mechanics of the human body. Doctors are selected for training largely on their ability to handle the intellectual rigors of such a complex task—a task that is proving ever more complex with each ensuing year.

In the early 1970s, at the time I was training to be a doctor, there was the sincere belief there would eventually be an effective drug for every known ailment. The pharmaceutical companies were gaining dominance, the genetic code was beginning to be understood, and computerized scanning was on the near horizon.

Patients were ill-informed. Those with cancer most often had their diagnosis hidden from them. Those with high blood pressure were not told their readings. As far as doctors were concerned, there really was no need to worry patients about such things. After all, that's why we were there! Although times have changed for the better with the age of informed consent, and Wikipedia, the legacy of those years remains.

Consultation times for family doctors are still short, not long enough to participate easily in the healing model I have just presented. Although it is changing for those better educated and more assertive in health matters, many patients remain conditioned to be passive partners in the management of their own health. The short time spent with their doctor is not enough for their doctor to act as an agent for change, challenging these limiting beliefs. There is often just enough time to classify the condition, write some notes, prescribe a medication, and sign a doctor's note for missing work.

All these tasks call for a focused mind ahead of an open heart, which is the reason why so many doctors, myself for one, have suffered burnout on the job. Young doctors who perceive this as their future feel discouraged and caught in a vise.

Over the past half century, pharmaceuticals have become the recognized tools of the medical profession. Rather than look deeper into the root causes of the health problem, it has been simpler and quicker to prescribe a chemical that controls the symptom. As doctors and health authorities have had to control their budgets, so pharmaceutical companies now promote their products directly to the public, many of whom are still conditioned to take a medication as an initial reaction to feeling unwell. So we, as a society, have come to regard our bodies as chemical, hormonal, and physical entities first, and energetic, spiritual beings very much second.

Every week we hear athletes talk about their latest adrenaline rush. We classify depressive illness as a chemical imbalance, and we are convinced that the reason we are light-headed is that our blood sugar is low. The medical model identifies linear chemical pathways in the body, and seeks to enhance or inhibit them, often using synthetic chemicals in the form of drugs. Such is the complexity of the body, however, that enhancing or inhibiting one pathway invariably impacts other pathways and systems, so other drugs are needed. And so on and so on.

In 2009, health care spending in the United States reached an estimated $2.5 trillion per annum, or $8,047 per person. By 2019, it is projected to grow to $4.5 trillion. Spending on prescription drugs in 2009 grew an estimated 5.2 percent to $246 billion, although this increase was partly due to the demand for antiviral drugs used to treat H1N1 ("swine") flu.[95] Despite the many advances in modern medicine, we have made little impact on our management of chronic illness. We have better immune-suppressing drugs now, but they come at the expense of serious side effects. For many conditions such as rheumatoid arthritis, lupus, multiple sclerosis, and chronic pain syndromes, the medical model has provided precious few breakthroughs.

When someone presents to a doctor in the early stages of a chronic disease, first and foremost their wish is for a diagnosis. They are keen for someone with medical training to make sense of all their strange feelings and sensations. In general, the family doctor diagnoses initially, then manages the condition to the best of her ability.

Sometimes, however, because she can't access the right tests or simply because the person's symptoms have not improved, she calls for help. The medical model encourages, quite rightly, early referral to experts. The doctor is seeking someone smart with more experience and time to spend thinking about the person's complex, seemingly insoluble problem. The patient then arrives at the expert's office, whereupon the cycle is repeated. It may be that the specialist finds a drug or an approach that the referring doctor has not considered, and a significant improvement results. (See Figure 37.)

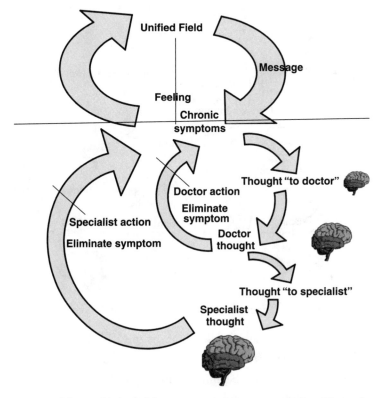

Figure 37. The medical model—patient to doctor to specialist. *The author stresses that the differences in brain sizes shown here are purely metaphorical!*

Unfortunately, by the time someone reaches my office, this successful outcome has yet to be reached. Or at the best, it has only resulted in temporary relief. More often than not, the person has seen many specialists, physical therapists, acupuncturists, and nutritionists, and greets me with the inevitable: "Help me—I seem to be going around in circles." (See figure 38.)

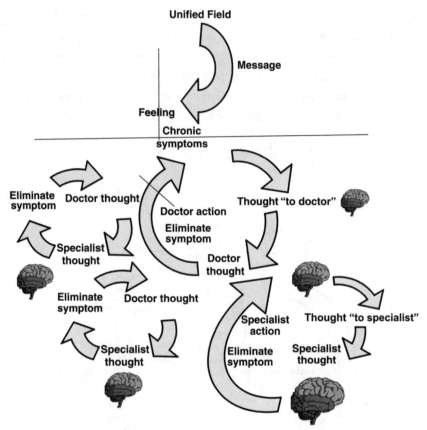

Figure 38. The medical model—going around in circles.

At this point, I explain that my brain is most likely no bigger and no more efficient than any of the others they have consulted. I venture to suggest that we need to start from the beginning and to hear their story—the way they want to tell it. We need to explore their passions alongside their problems; their triumphs alongside their traumas. Their childhood, their adolescence, their first love, and their last love. In place of brainpower, we have time to listen. Time to involve them, and time to simplify their life. So we embark on the healing cycles as illustrated in figures 34 and 35.

At present, our formal medical model does not embrace the concept that other dimensions exist beyond those we perceive with our senses. Consciousness is simply a word to describe the state we are in when awake, or safely out of the hands of our anaesthetist colleagues. The quantum world is something found in science fiction and James Bond movies, and has little relevance to the realities of getting people better. Its effects (the Big Bang is presumably an exception) are probably minimal. According to this model, most illness has no intrinsic meaning apart from the obvious need to stop smoking and eating junk food. We should firmly stick to our knitting and not venture into the vague, flaky world of metaphysics—leave all that nonsense to the fortune-tellers and snake-oil salesmen.

Well, maybe I am being disingenuous. In my many discussions with my medical colleagues about this work, the most common reaction is one of restrained interest. The medical system has made them weary, and it is altogether too hard and draining a prospect to embrace a new paradigm. The structure of medical practice does not allow them the luxury of such a style of practice, and many changes need to be implemented before this approach becomes commonplace. In this book, I hope to present the case that it is not an absence of scientific proof that is holding up this advance. In fact, I believe we are at the threshold of a revolution in science that will validate at least some of the theories that support the notion that we are holographic beings. Instead, we are being held back by a structure that is ill-suited to our needs.

What is needed is the energy and vision to support the emerging paradigm. It is likely that educated, informed people will lead this change, and that a medical model owned by us all will respond, adapt, and expand, accordingly. It is clear that, as doctors, we need your help.

Chapter 19:
Cocreation and Free Will

All dreams spin out from the same web.

—Hopi proverb

There are many interpretations of the famous petroglyph etched into the sandstone cliff, now known as Prophecy Rock, on the Hopi reservation in Arizona. Many feel it depicts two possible courses open to mankind. First, a destructive course governed by humans each possessing two divided hearts—a metaphor for those who think only with their heads, separated from their "feeling hearts," in a deductive, highly rational way. And second, a sustainable future led by those each with one heart—a metaphor for those with heart and brain working together in a state of compassion for the world and all its inhabitants. The two-hearted humans are given the opportunity to unite their hearts, and achieve salvation for themselves and their planet.

Over the years, I have observed the many simple health benefits accrued by those who allow their hearts and minds to integrate in this fashion. It is perhaps for this reason, above any other, that I have been attracted to the time-honored philosophies of traditional cultures. For example, the Vedic understanding of the growth of human consciousness, as illustrated by our ascent through the chakras, appeals as it clearly demonstrates that our heart must be engaged in compassion for others, and ourselves, if we are to achieve balance and joy in our lives. With our heart engaged, we can then progress to forge our true way, our vocation, on this earth, balancing our intellect with our intuition. With this balance achieved, we become portals, or antennas, for endless creative ideas, and become refreshed and rejuvenated each time we convert these ideas into actions and events that make a difference. If I am to interpret the prophecy of the Hopi correctly, it is essential that we evolve into such integrated creative beings if we are to survive. Figure 39 depicts our whole body as the receiver of true inspiration—creative ideas that connect with us instantly as we live in the moment, at one with a unifying field.

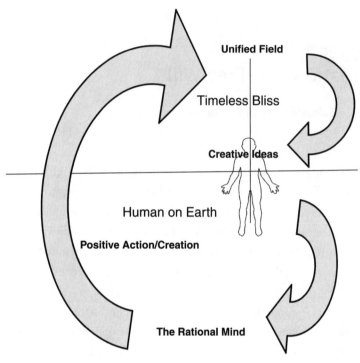

Figure 39. Cocreation.

As we progress in our life through the layers of consciousness, we are protected against those who wish to infect us with their conditioned, restricted thoughts. If we are lucky enough to have been nurtured compassionately through our childhood years, then we will have achieved the self-respect needed to challenge those trying to control us. Yet we will also have learned that it is unwise to try to control the mind-set of others—better to influence others through our own example. We will also have learned to study with an open mind, and to experience education as a process of "leading out" rather than of "forcing in." And if our childhood years have been less than perfect, then we understand that life's experiences help us evolve both emotionally and spiritually. We realize that we are continuously surrounded by teachers, in the form of our friends, our adversaries, our children, and our animals. And so from cradle to grave, we are perpetually learning. We are learning that we have an infinite array of choices every moment we are here. We are learning that our most potent force is our free will—a precious gift that, if used with wisdom, carries with it the power to create a magnificent future.

For this reason, above all others, it is important to regard prophecies less as sinister omens dictating our future, and more as guides helping us to proceed with wisdom. Their words and images may well be set in stone, but their predictions are not. They are not there to scare us, nor to make us feel smug. Rather they may serve as an astute warning, or as a reassuring confirmation that we are on the right track.

The Mayan prophecy indicating tumultuous changes at our particular time in history is a case in point. There will be those who study the Mayan calendar and interpret it as ending on December 21, 2012, and be fearful that on this precise day our time on Earth will come to an abrupt end or, at the very least, we will all face an overwhelming catastrophe. And there will be those who interpret it more liberally as a metaphor—that this is a time of great change and growth of our consciousness, and that we indeed have to make important choices and follow up on them if we are to thrive and survive.

Now, more than ever, is a time to allow creative ideas to flourish. There is no doubt that the imminent threat posed by climate change demands highly creative yet practical solutions. Lateral thinking is now essential. For example, selecting one recent example from many, there is the inspired idea that our cars can be fueled, not by nonrenewable petrol but by sustainable human waste. Enterprising engineers at Wessex Water in Bristol, England, have produced a successful working model of a Bio-bug, a converted Volkswagen known affectionately to locals as the Dung Beetle. Bacteria break down human waste to form the methane gas that powers the car. Although as a result, it releases carbon dioxide into the air, it is carbon neutral, as this would have been released into the atmosphere anyway in the form of methane.

There are many such creative projects flourishing in communities around the world; each bicycle lane, walking track, and recycling scheme began life as one person's inspired idea, shared with colleagues then followed through with passion, persistence, and know-how.

Chapter 20: Cults and Control

Thoughts are like arrows: once released, they strike their mark.
Guard them well or one day you may be your own victim.

—Hopi proverb

As a young boy, Adolf Hitler loved to play cowboys and Indians. His favorite cowboy character was Old Shatterhand, the hero of numerous novels by the famous German author Karl May (who had never actually set foot in America). The stories are narrated by Old Shatterhand, who forms a close friendship with the fictional Apache chief Winnetou, and together they experience the high adventure of combat while developing a deep compassion for nature and their fellow man. In the course of his life, including the time he was Führer, Hitler avidly read and reread as many as seventy of May's novels. He even encouraged his officers fighting on the Russian front to study their contents, and to pay particular attention to Winnetou's "tactical finesse and circumspection." By this time, however, many of Karl May's novels had been reedited in an anti-Semitic style, completely at odds with the author's original humanitarian intent.

In 1900, when Hitler was ten, his younger brother died of measles. As a result, he changed from being a happy and playful child into one who was sullen and detached. He clashed continuously with his authoritarian father who frequently beat him. The young Hitler's bitterness and resentment grew deeper when his father forbade him to go to a school devoted to his first passion, art. His father died when Hitler was thirteen, leaving these conflicts unresolved.

None of these insights into Adolf Hitler's early life explain why he grew into, without doubt, the most dangerous and murderous man in history. A man capable of hypnotizing a nation into becoming blind to their better natures, and into relinquishing any responsibility for their actions. After all, at the end of the nineteenth century, strict fathers, dying siblings, and shattered childhood dreams were very much the order of the day. There is no room, now or in the future, for excuses.

We are left pondering, however, how different it might have been had Hitler's early years proved to be less confrontational. World War II, Hitler's legacy, left sixty million people dead. An estimated eleven million people, more than half of whom were Jewish, died in the holocaust at the hands of human beings convinced that the extermination of those from a different race and of those holding different philosophies was ultimately a gift for humanity. For the premeditated actions of those war criminals whose minds were controlled in this way, there too can never be any excuses.

It is clear, however, that we who belong to succeeding generations have a duty to learn from our devastating recent past. As adults, it is our responsibility to protect ourselves from those who wish to plant ideas into our heads. We must be allowed to think for ourselves and challenge dogma where necessary. And we must protect our children.

There is now convincing evidence that children exposed to violence on television, in movies, and in computer games are at risk of behaving aggressively both in the short-term and the long-term. In 2007, researchers from the University of Michigan identified the risk of emotional damage to young people from violent programs to be second only, in statistical terms, to the risk of lung cancer from smoking.[96] A child's mind is open, her feelings exposed. An EEG would commonly show an alpha or theta brain-wave pattern, ideal for creative play but leaving the child vulnerable to the power of suggestion. Ideas can be lodged deeply in the child's subconscious, as she has yet to develop the skills needed to discriminate, or to rationalize her feelings.

Adults too are susceptible to the process of mind control, commonly labeled *brainwashing*. One of the pioneers in this field is the American psychiatrist Robert Jay Lifton, who prefers yet another term, *thought reform*. Lifton studied U.S. servicemen who had been held captive in the Korean War. In his 1961 book, *Thought Reform and the Psychology of Totalism: A Study of "Brainwashing" in China,* he analyzed the various methods of thought reform employed at the time in Eastern communist states. He identified eight salient points aimed at helping people decide whether a process of mind control is being enacted. I have added a brief explanation of each point:

1. **Environment control.** This limits communication with the outside world.
2. **Mystical manipulation.** Claims of a higher purpose or demonstration of a "miracle" early in the process are signs of such manipulation.
3. **Demand for purity.** The group alone is responsible for significant change in the person and in society as a whole.
4. **Cult of confession.** The insistence of open disclosure of past "sins" as well as negative thoughts about the group.
5. **Sacred science.** The group's perspective represents the absolute truth; there is no room for robust debate.
6. **Loaded language.** Strict adherence to a limited or altered language; this restricts debate and limits discussion to "black-and-white" dogma.
7. **Doctrine over person.** Personal experiences that conflict with the doctrine are denied.
8. **Dispensing of existence.** Those who leave or exist outside the group are doomed.

In a later book, *The Nazi Doctors,* Lifton examines how and why members of the medical profession were able to supervise mass murder in the Nazi concentration camps. It is a chilling investigation of the underlying psychological mind-set that drove intelligent, vocationally trained doctors into justifying and rationalizing their role as "killer-healers." The reasons are complex and still difficult for us to fathom. However, Lifton identifies that the doctors acted this way, at least in part, in an attempt to overcome a profound feeling of powerlessness.

Hypnotic techniques, with or without specific medication, can also be used to induce a state of mind control. In a 1995 *Wall Street Journal* article, five hundred people a month in Colombia were reported to be the unwitting victims of poisoning by the drug Burundanga (the anti–motion sick-ness pharmaceutical scopolamine is refined from this compound). Classically, someone would ingest a drink laced with this poison by a street criminal, and immediately lose their free will. They would hand over money and jewelry, and even make bank withdrawals in the robber's presence. When the effects of the drug wore off, the victim would retain no memory of the event.[97]

Throughout the developed world, Rohypnol, a sedative and hypnotic medicine from the benzodiazepine class, has gained a reputation as the date rape drug. It has commonly been used in combination with another drug, GHB (gamma- hydroxybutyrate), to spike the alcohol drink of a potential victim. The spiker would then hope the victim would lose her will to say no, and comply to having sexual intercourse. She would subsequently have no memory of the event. Recent studies suggest, however, that the most common date rape drug is alcohol itself.[98]

Adults can also succumb to more subtle, yet perhaps just as insidious hypnotic techniques that do not rely on the ingestion of drugs or alcohol. Over the years, I have attended many workshops on healing techniques, and the vast majority are ethically and skilfully presented. As we have already discussed, healing practices invariably employ techniques aimed at inducing a state of deep relaxation in their clients, and so healers are in general very skilled at this art.

I have at times witnessed a facilitator of a healing workshop induce such a state in his attendees, and then perform a dramatic "miracle" on someone from the group. This state of focused relaxation is the ideal environment in which to demonstrate a healing act, for example, the instant resolution of a frozen shoulder. The relaxed state of the healee, and of the audience as a whole, contributed to the "healing field." Of course, all this could be deemed beneficial, especially for the healee. In an ethical environment, the process would be explained, and the roles of all the contributors, including the audience members, duly honored. It would be pointed out that none of this was due to any special powers inherent in any of those present, and that others using similar techniques can produce a similarly pleasing result.

There are times, however, when such a scenario is not played out. I have learned to be wary if, once the attendees are induced into a relaxed state, the workshop leader discredits other recognized healing techniques. I am wary too if the workshop leader proclaims that a person not responding to a healing "really doesn't want to get better." And I become particularly wary of the suggestion that further progress in my healing journey will only happen if I sign up, there and then, for a more advanced course.

Of course, ultimately we are all responsible for precisely how and what we choose to learn. There is good evidence that many of the long-term, sinister effects of mind-control strategies do eventually wear off. I have seen many middle-aged folks who spent years of their youth in closed cults who now think and act freely. In some cases, their children have been their salvation, teaching them home truths in the way only our children can. I believe the greatest protection against mind control we can offer ourselves is an ever-present awareness of our right to exercise our free will. And as the Hopi advised, we should not only be cautious of being struck down by other people's malicious thoughts, when, like barbed arrows, they are fired in our direction, we should also guard against those we aim, recklessly, at ourselves.

As we progress in life, we mellow. We learn to be open-minded, yet seldom gullible; healthily skeptical, yet seldom cynical. We learn to be wary of any extremist who expounds dogma, and to be protective of our children, and of the vulnerable, until the time comes when we perceive they are ready to be truly independent.

Summary of the Experience of the Human Hologram

The human hologram model recognizes that beneath and beyond our universe's physical structure, and beyond our understanding of energy, lies the realm of pure information. Our bodies detect this information, and indeed are formed upon its matrix.

There is no hierarchical order to the human hologram; its organs, blood, and connective tissue, its bones, muscles, and tissues combine and complement each other in a truly integrated fashion. In section three, however, I focused on the role of two vital organs: the heart and the brain.

The heart is addressed first, as there is evidence that along with its physical role as a pump distributing oxygen and nutrients throughout our body, it is also our most sensitive detector of information. The heart plays a role in detecting subtle changes in this field of information, and in relaying these changes to the rest of the body, including the brain. We examined the evidence that this information is capable of being transmitted beyond the confines of our physical bodies, to others near and far.

Encoded into this information are our feelings; feelings play a prominent and primary role in the human hologram model, confirming to us our existence and indicating to us the need for change when appropriate.

The brain is duly honored as our organ of perception. Our brain processes the information received from our senses, and transforms it into the familiar dimensions of time and space. We are here, I surmise, to act, and we need a viable framework within which to perform.

We turned to the pioneering theories of Karl Pribram, Sir Roger Penrose, and Stuart Hameroff to examine just how this transformation might take place. Pribram's work theorizes that our senses act as special lenses, bringing into focus underlying fields of information, transforming them (through the Fourier model) into a usable format, ready to be fully interpreted by our brain cells. Building on this, Penrose and Hameroff's work theorizes that it is within the microtubules of our brain cells that this act of interpretation takes place. This is where our perception, or our consciousness, is processed, where we make sense of the quantum world. Or, as physicists would say, where the quantum world collapses.

The human hologram model honors our rational brain alongside our intuition. Intuition, though, requires a special partnership with our feeling heart, and is best experienced when our rational mind is quieted. For humans to live to their full potential, a balance needs to be achieved between feeling and thinking. In early childhood, we are sensitive and vulnerable. It is our feelings rather than our rationality that predominates. It is highly important that our parents or guardian honor this, protecting and guiding us with sensitivity and compassion.

As we grow, we need role models who themselves have learned to find this ideal head-and-heart partnership. With their help, we learn to relate, rather than simply to react to our feelings, and to resolve issues as they present to us. We learn to value our free will, and to use it wisely. These skills contribute directly to our future physical and emotional health.

The fifth guiding principle of the human hologram explains how each of us, through our own observation, participates in our world and our universe. There is an ever-present subtle partnership between light, matter, and ourselves. The subject and the object are inseparable. And, so woven into section three, are my own experiences and perceptions. In my working life, I continue to enjoy the privilege of meeting many people at times of personal crisis. These are the times that honesty tends to prevail.

These experiences have led me to observe that the process of deep healing involves embarking on a new cycle of awareness, often learning for the first time how to be compassionate to ourselves. Just how we achieve this requires us to be patient and smart in equal amounts. If I am asked to help another, it is as if I am, for a short time, required to be a tolerant and smart parent. Sometimes this works well; other times, it is far from perfect. It is, after all, simply a human endeavor.

As each healing cycle is completed, others benefit. The compounding effect of this is no less than the sustainable growth of human consciousness for generations to come. This, for me, makes the human hologram model highly appealing.

Section Four
The Human Hologram—The Speculation

Introduction to the Speculation

It is fine to speculate. All theories are speculative, and it is important as an exercise to imagine just where our theories might take us. In addition to being immensely exciting, it is also the responsible thing to do. As doctors, we only make an attempt at reaching a *diagnosis,* so that we can then move on to the most important piece of information, the *prognosis.* First we analyze the data, then we decide, through speculation, whether someone's future is either bright or bleak.

And so it is pertinent to ask ourselves if, indeed, we are holographic beings, just what does this mean for us? Is it, indeed, something that we need to know? Or is it, after all is said and done, an exercise in futility, flawed in concept, ultimately leading us down a road to nowhere?

With this in mind, I now invite you to join me in this exercise in purposeful prognostication. For my part, I'll concentrate on those subjects about which I have, primarily through my longevity, gained some experience. I encourage you to apply your hard-earned wisdom to edit or to expand my vision. There is so much more to learn.

Chapter 21:
The Human Hologram—Death and Life

Death is for many of us the gate of hell; but we are inside on the way out, not outside on the way in.
—George Bernard Shaw (1856–1950), Irish playwright and author,
A Treatise on Parents and Children

Our now familiar human hologram model, and especially the far-reaching concept of the biohologram, poses many questions about the true nature of our existence. Are there dimensions that lie hidden from us? Is everything we see, hear, smell, and touch a mere perception, a virtual reality? Is our physical body part of this virtual reality?

By raising these questions about the fundamental nature our life here on Earth, we unavoidably present ourselves with another equally profound dilemma. What happens to the human hologram when the physical body dies?

It can certainly be said that there are magical moments that occur in all our lives that could justly be described as heavenly. The moments of bliss where we rest in peace; the moments of spontaneous joy when we laugh with another. According to the human hologram model, these are the times that we are at one with the unified field. Or as one could interpret the Lord's Prayer, when we experience God's will as being done "on earth, as it is in heaven."

And equally, we can all recall times that have been far from heavenly. Times when we have felt powerless in the face of aggression or abuse. Times of chaos, pressure, and extreme frustration. In the real-life stories recounted in this book, there are many such pivotal events that could aptly be termed hellish.

One patient visiting me last week described how he had witnessed his wife being murdered in Africa in front of his children. For so many, the tragic legacy of conflict and past wars persists relentlessly day after day. For those who suffer in this way, George Bernard Shaw's vision of death as a welcome escape route from hell would not seem so far-fetched. Figure 40 shows how destructive acts threaten the potential of humans to create heaven on earth.

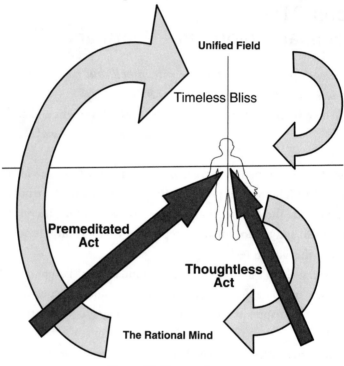

Figure 40. Destructive acts.

These destructive acts may be due to ignorance or be premeditated. Poor listening skills are more prominent in those lacking an empathetic upbringing or a sound education. The person reacting to his feelings in an angry or aggressive manner is likely too to have had poor role models in his early life. At times of deprivation, he is then likely to lash out with his tongue, his fists, or, worse still, a weapon. Our prisons are packed with poorly educated criminals who have committed violent crimes of passion, often fueled by dangerous addictive chemicals. Premeditated planned destruction spearheaded by narcissistic and intelligent humans can prove even more dangerous. Cults form, attracting the lost, the downtrodden, and the ignorant, and, unless we are wary, the oppression can spread like a cancer throughout society and last for generations to come. As the stories in this book relate, however, with time and compassionate support, healing does ensue, lives reform, and the darkness slowly fades. Education is the key.

And our education should include a healthy grounding in the subject of death. It is understandable that a free discussion of death has remained a taboo in the West. None of us likes to embark on a discussion on something about which we know so little. As the subject of death appears to deal with the ultimate loss, that of our very existence, it is likely to produce discomfort in some, and expose unresolved feelings of grief and guilt in others. Our focus in Western society on material wealth, material health, and the material sciences has not helped open or fuel meaningful debate on the topic. So conditioned are we by the constraints of this paradigm that even in highly articulate, educated circles, the subject of death is something seldom embraced. Maybe for those at the top of their game, or at the peak of their intellect, their own demise is just too painful to contemplate. Death is something from which none of us can remain aloof.

For the purely rational amongst us, it would seem logical to assume that our death will simply mean that we cease to exist. And, with this mind-set, it would be tempting to regard those who, through their religious faith, believe that an afterlife beckons as being naïve in the extreme, wishful thinkers at best.

The advances in our understanding of physics over the past century lead us, however, to question such a dogmatic stance. In March 1955, Michele Besso, the engineer and close friend of Albert Einstein since their days at the Federal Polytechnic Institute in Zurich half a century earlier, died at the age of eighty-one. At his friend's funeral, Einstein said: "Now he has departed from this strange world a little ahead of me. That means nothing. People like us, who believe in physics, know that the distinction between past, present, and future is only a stubbornly persistent illusion." Albert Einstein died one month later.

Modern science has now ventured outside the narrow parameters of time and space, thereby giving us a wider perspective on our lives—and our deaths. Energy and matter we know to be interchangeable; time is relative. According to the first law of thermodynamics, energy—this includes *our* energy—cannot be destroyed, only recycled. And according to the second law, complex physical structures—our bodies, for instance—inevitably decay.

Quantum physics acknowledges that other realms exist, and that our physical presence depends on them. The Copenhagen interpretation of the uncertainty principle allows us to consider that we play an integral part as observers. And the holographic model of Bekenstein, Susskind, and 't Hooft forwards the idea that a matrix of information is fundamental in our universe, and that the reality we perceive is a construct of our minds.

So is science leading us to believe that the only reality we lose when we die is virtual? If this is the case, do we continue to feel? Or do we return to our source—scattered into a vast sea of consciousness—safe and sound, but mute and oblivious? Unfortunately, science as yet cannot supply us with any hard facts to help us answer these questions. Scientists who, like the rest of us, would prefer the option of eternal bliss must rely on faith alone. "And to the faithful, death the gate of life," wrote John Milton in his epic poem *Paradise Lost.*

Our own most valid experiences of death are intensely personal and deeply felt. When observing the body of someone close to us who has recently died, it is clear that the person we knew so well has gone. We may somehow feel his presence, but know that he no longer resides in the body before us. He has left behind the body that touched, saw, heard, and smelled others. The heart that relayed energy to his body and beyond has ceased to beat. The complex computer we called his brain, which coordinated the complex workings of his body, which perceived and made sense of its surroundings, functions no more. Our loss is that we can no longer enjoy the physical company of a friend; his loss is that he can no longer engage in our world. For him, it is his universe of time and space that dies.

Of course, none of us knows for sure what happens after death. Between 10 and 20 percent of those who go through cardiac arrest and clinical death and are then successfully resuscitated report near-death experiences (NDEs). In addition to experiencing a detached awareness of their physical surroundings and the frantic goings-on in the emergency room, they commonly describe a state of bliss similar to, but far more profound than those found in our heavenly moments here on Earth—a sense of release, levity, and love. A state devoid of fear, guilt, worry, or anger. A tunnel, a light, a benign presence, a feeling of coming home. For some too, an instant life review.

In his book *Glimpses of Eternity,* one of the pioneers in the field of NDEs, Dr. Raymond Moody, describes how such experiences can be shared by loved ones at the bedside of someone dying.[99] I have heard many similar stories in my time as a doctor—something that appears very natural at a time when hearts are so open and the veil between dimensions has been lifted.

At the time of writing, the world's largest ever scientific study of near-death experiences is under way, under the guidance of Dr. Sam Parnia at England's Southampton University. Known as the AWARE study (AWAreness during REsuscitation), large number of patients in the United Kingdom, Europe, and North America are being studied over three years, using sophisticated brain and body monitoring techniques. The researchers will also test the validity of whether, and exactly what, people can see and hear during the process of resuscitation. They aim to determine whether a state of awareness persists even when the common markers of clinical death—cardiac arrest, absent brain function, and other physiological data—are present.[100]

This is important research into a deeper understanding of the nature of human consciousness. It will help determine whether our consciousness resides, in some form or other, beyond the physical constraints of our body. If the evidence supports this, then the possibility of consciousness surviving physical death can begin to be entertained by the scientific community.

As important as an understanding and a validation of NDEs is, however, it is merely a first step taken in a more challenging exploration of the scientific nature of death itself. Perhaps it can be envisaged as a taster, a tempting hors d'oeuvre to a main course whose ingredients no one, not even the surviving ND experiencers, can begin to guess.

But if there is one truth that already shines through the NDE research, it is that those who have lived through such experiences consistently report that they no longer fear the actual process of dying. It would appear that fearing death is not good for us. In the 1980s, Dr. David Spiegel conducted a landmark study at Stanford University into the beneficial impact of support groups on patients with advanced breast cancer.[101] The sessions included free discussion on ways the women could face and master their fears of dying. The initial aim of the support groups was to help patients manage their symptoms and the side effects of their treatment. When records were studied ten years later, however, it was discovered that those attending the groups had survived significantly longer.

It would seem prudent and more proactive to address our fears around death and dying early in life. My grandmother Olive, a progressive, no-nonsense thinker who opened Britain's first Montessori school shortly after World War I, was keen for each child to have a pet. The pet would bring the child great joy, but needed to be cared for, thereby teaching the child discipline, patience, and future parenting skills. But moreover, when the pet's time came, the child would be introduced to the concept of death—the sadness, the practicalities, and the natural order. Certainly, this would provide the child with a profound and precious memory to be held and valued as the child grew to adulthood; so much healthier than disturbing images of death imprinted in a child's subconscious as a result of watching violent movies or playing violent computer games.

The Scottish born conservationist John Muir, who through his activism helped establish the Yosemite National Park in California, understood that the natural world would provide children with their best education, helping dispel any unhealthy fears of dying that could cloud their enjoyment of living. He stated: "Let children walk with Nature, let them see the beautiful blendings and communions of death and life, their joyous inseparable unity, as taught in woods and meadows, plains and mountains and streams of our blessed star, and they will learn that death is stingless indeed, and as beautiful as life."

Modern technology is allowing us to explore the natural world in greater detail than ever before. We are now able to see nature close-up, observing cells, molecules, atoms, and beyond. We are even able, armed with a femtosecond laser, to witness the presence of nature's "joyous inseparable unity" in the form of quantum biology. The human mind, a natural wonder of the world, is capable of understanding the mathematics underlying our universe. In partnership with that greatest of all modern inventions, the computer, it can even begin to untangle the previously hidden order and chaos of nature through an understanding of fractal geometry.

The beauty of life in no way seems to be lost as we explore beyond where the naked eye can see. Somehow too, as science reveals the existence of other dimensions, death itself does not appear as sinister nor as dark. For Helen Keller, deaf and blind since an illness at the age of nineteen months, death held little fear. "Death is no more than passing from one room into another," she once said. "But there's a difference for me, you know. Because in that other room I shall be able to see."

None of us knows what will ultimately reveal itself when we step out of this very cluttered "living" room. Away from all the distractions and chaos of our lives, will other dimensions show forth?

Time—if time for us were then to exist—will surely tell.

Chapter 22:
The Human Hologram—The Future of Science, Medicine, and Technology

There will still be things that machines cannot do. They will not produce great art or great literature or great philosophy; they will not be able to discover the secret springs of happiness in the human heart; they will know nothing of love and friendship.
—Bertrand Russell (1872–1970), British philosopher

There are those who predict that, over the next four decades, technology will advance to a point that human intelligence will be outperformed and overtaken by artificial intelligence (AI). The predictions have been made by plotting the exponential growth of computer power and information technology over the past three decades, and extrapolating this into the future. In medicine, so it is said, the ever-more sophisticated computerized imaging techniques and the widespread use of nanorobots delivering drugs and radiation selectively to our diseased cells will result in us all enjoying longer and healthier lives. Nanorobots will be attached to our brain cells, making us incredibly smart. All this will lead to an inevitable moment in time, labeled "the singularity point," when AI and humankind combine to form a future beyond our wildest (present) imaginings.

It is clear to all that computer technology is advancing at a breakneck speed. It is also apparent that the paradigm will shift once quantum computers become viable. Futurist Ray Kurzweil has been one of the leading proponents of this technological vision of the future. In his famous 2001 essay, "The Law of Accelerating Returns," he extended Moore's law, which describes specifically the exponential growth in semiconductor efficiency and information storage capability in computers, to cover the rapid compounding growth of information and nanotechnology as a whole. His recent book *The Singularity Is Near,* predicts that, by 2018, computers will match the storage capacity of the human brain; by the 2030s, nanomachines planted in our brains will be able to produce a complete virtual reality for us; and by 2045, the singularity point, we will be able to buy for a thousand dollars a computer a billion times more powerful than our own brain. Along the way, nanotechnology will be used to "enhance" the minds of humans, and to make our body's physiology more efficient. This means that by 2045—and this is cold comfort—the machines are unlikely to have taken complete control, as humans themselves, the smart ones at least, will have become human-machine hybrids. After this date, however, there are likely to be more and more tipping points—"aha" moments for the machines—as this exponential growth in smartness leads to an explosion of ideas and an uncovering of new paradigms, allowing new versions of humans to expand their role here on Earth and throughout the cosmos.[102] And for those who are wary of these advances, for those who dare to resist, what is their fate? Well, some have already labeled them, in advance, "the new Luddites."

Legend has it that in 1779, a wayward British youth named Ned Ludd, on being accused of idleness, became so enraged that he smashed two knitting frames with a hammer. Over the next century, those activists who despaired that machines were taking over from skilled workers in the Industrial Revolution subsequently took the name Luddites. The Luddites of the twenty-first century, so the prediction goes, will actively resist the forced imposition of technology on and within them and others. They will view it as a dangerous assault on nature and on true human values. Ultimately, though, their campaign will fail as they become outwitted and outnumbered. Powerless, they will be confined to live as outcasts, like animals on specific nature reserves.

The human hologram hypothesis challenges this, perhaps extreme, vision of the future. It does so on many levels. It prefers to view the undoubted technological advances that will greet us over the next few decades as extensions of our own evolution. Technology that must be used with great integrity. Indeed, the intelligence of the human hologram is far more profound than the calculating, information-processing ability of a computer. Or even a computer that convinces itself it is more powerful than one billion brains. What is more, if the microtubules found in human brain cells are truly the sites for quantum processing, then, by Dr. Stuart Hameroff's calculations, our brain is capable of 10^{28} operations per second, a trillion times faster than originally thought. If this is so, it is likely that any singularity point will be delayed for several decades—or perhaps forever.[103]

The alternative vision of the future does not take an extremist Luddite stance. Indeed, nanotechnology will allow much of medicine to become less invasive, safer, and more focused. But the microscopic machines injected into our bloodstream to help replace damaged cells or to radiate cancer cells are still likely to be no more than tiny ambulances hovering at the bottom of a very foreboding cliff. As today, the metaphorical cliff will represent the real causes of human disease—poor nutrition, our toxic environment, unresolved emotional issues, and those issues yet to be discovered. There are, of course, inherited diseases that will benefit greatly from advanced gene therapy techniques. But in many cases, these benefits will not prove sustainable if epigenetic and environmental factors are ignored.

Over the years, the most astounding medical breakthroughs have resulted not from technology, but rather from a harmonious blend of many fine human traits. It cannot be denied that intelligent reasoning is one of these, but in most instances there has also been a healthy mix of good intent, serendipity, and passion.

On the day of writing, the world has lost a doctor and researcher whose work has saved thousands of lives. In the late 1960s, Professor Sir Graham Liggins became convinced that babies could be prevented from being born so prematurely that they failed to survive. With this goal in mind, he conducted meticulous research on pregnant ewes undergoing premature labor, injecting them with steroids in an attempt to prolong their pregnancies. This did not happen. Instead, the lambs continued to be born prematurely, but then survived, and thrived, outside the womb.

He then proceeded to trial this therapy on humans. In 1972, his landmark paper was published in the journal *Pediatrics* (having been rejected by the prestigious *Lancet*), and since this time it has become common practice to inject mothers going into early premature labor with steroids.[104] Prior to this, most babies born before thirty-three weeks' gestation died, as their lungs were not mature enough to cope. The steroids help the baby's lungs to mature quickly.

There are thousands of people alive today solely because of Professor Liggins's work. Maybe I'm influenced by the kindness shown to me by this very humble man when I was a very junior doctor, but I doubt that the humanity, intelligence, and integrity he possessed could ever be enhanced, or matched, by any artificial machine.

This brings us to the core of the difference. The human hologram model acknowledges that it is our feelings and our experiences that are of fundamental importance. It recognizes that our bodies work together with our brains in a truly integrated fashion. The body feels, the brain thinks. Insights and inspirations come from a coherence throughout the body, a balance of head and heart.

The human hologram model further defines our feelings and our symptoms as subtle variations within a shared universal field of consciousness, detected by our bodies, and by our hearts in particular. It proposes that these are fundamental messages more pure, more direct, and more powerful than we have been conditioned to realize. Whereas the singularity model seeks to create exciting virtual realities for us via nanocomputers lodged in our brain, the human hologram model suggests that, through our senses and brains, we are already creating our own virtual reality—these dimensions of time and space. And that this perception is important for us to do our work here within these very dimensions. It gives us a code to live by.

I would hope that this, the more expanded scientific view of the human condition, empowers health professionals to rediscover the art of listening, and encourages them to employ their innate compassion as a vital healing tool. Of course, not all will be drawn to follow this challenging vocational path of philosopher-healer. If the vital importance of this approach is honored and instituted alongside the necessary academic training of health professionals and doctors, however, then I suspect a new level of health care will be reached.

Modern technology can then find its rightful place, employed for the greater good. For example:

1. High-speed Internet to reach all areas in the world, helping with relief programs and helping to bring people together.

2. Diverse educational programs that empower children to reach their full potential.

3. Medical screening techniques that are simple enough to be used by the public at large.

4.   Less invasive medical treatments targeted to heal without side effects.

The coming decades will also see the coming-of-age of the recreational and traveling human hologram—real-time 3D images of us appearing at distant locations of our choosing. As this science develops, and we become accustomed to welcoming our friends visiting us in holographic form from the other side of the world, it will dawn on us that the advanced technology that creates this illusion is but a pale imitation of something we already possess. We will begin to marvel at how our eyes and our brains create the holographic visions we see around us. The more we understand about photons, polarity, and interference patterns, the more we will understand about ourselves.

We will strive to understand more about our own living virtual reality before overindulging in those created by others as an amusement or a diversion. And we will seek to understand what is happening deep within us, so that appropriate internal changes are attended to first and foremost. We will realize that the only way for humanity to evolve is deeply personal, a process that unfolds within each human being. As this happens, no unfeeling AI or human-machine hybrid can compare or compete, whatever the size of their hard drive.

The twentieth century introduced us to the mysteries and the realities of quantum physics. Much of this new science remained hypothetical until this century. Over the past decade, scientists aided by cutting-edge technology have discovered that the quantum world exists not only in cold dry labs, but also in the warm environment of living beings. Whereas ten years ago, those who posited that we as observers could influence our reality were regarded as gullible New Agers, serious vocational scientists cannot now escape the fact that their own consciousness can influence the outcome of their experiments. It seems whenever we are drawn to examine in the most minute detail the workings of the world "out there," we discover just as much about the secret world within. The twin slit quantum eraser experiment of 2002 (see appendix 2a) is one example that sheds light on the nonlocal reality of the quantum world, of which we are very much a part. In this study, changes in one beam of entangled light occur ahead of the time the experimenter interferes with the polarity of the other beam.

In 2007, French physicists at the École Normale Supérieure de Cachan performed an equally impressive experiment devised over thirty years ago by the renowned physicist John Wheeler; however, it is only in this decade that we have had the technology to make it happen.[105] The experiment showed that photons "decide" whether they are going to behave like bullets or waves in response to a later decision made by a random event generator, something outside the immediate conscious control of the scientists. This bizarre result is making scientists ponder even more deeply about the nature of light, and our relationship with this energy that is so essential to our survival. It appears that we are getting close to an understanding of the behavior of light in its quantum state, even before we subject it to our gaze. Could it be that it is the *underlying intent* of the experimenters that is more important than their *immediate visual observation,* and that this intent somehow transcends the barriers of time? (See Appendix 2b.) Or does the answer lie deep within the hidden world of the photon itself?

This twenty-first-century research seems to be confirming that fields of energy exist in nature outside our accustomed parameters of time and space. Chapter 15 examined the possible ways our brains interpret the information (or light energy) reaching our eyes, transforming it into a code that is eventually unraveled to create images of our world. For this to happen, there has to be an interplay between photons and our own cells, so that we can make some sense of the bizarre quantum world of light.

Our special relationship with the quantum world also reveals itself through our sense of smell. Chapter 7 explored the evidence that the unique fragrance of a perfume represents a state of entanglement between the smell and the person who smells. Although studies relating to our other senses are yet to be devised, one need only observe the impassioned performance of a concert pianist at the height of her powers, to know how beautiful sounds can infuse a human being and radiate out to others, enraptured together within a timeless space. It is at a time, outside time, like this that we genuinely experience a slice of heaven on our earth. There are many who would extend this line of reasoning to their sense of taste, as they sample the divine delights of a Black Forest gateau or savor on their palate the subtle texture of a pinot noir wine from a boutique vineyard on the South Island of New Zealand.

My own professional interaction with the holographic paradigm is perhaps not quite so exotic. My experience tells me, however, that a formal acceptance of ear acupuncture, and other simple-to-learn therapies based on holographic principles, as an effective primary care tool will lead to a significant reduction of human suffering. The need for potentially harmful and expensive medication could also be minimized.

But there is little doubt that the most remarkable scientific application of the holographic paradigm has been the work of the Russian physicist Dr. Peter Gariaev. His wave genetics research honors the vision of many twentieth-century and contemporary scientists—Gabor, Wheeler, Bekenstein, Susskind, 't Hooft, and Mandelbrot, to name just a few. It is a logical extrapolation of the holographic universe theory that our bodies too must be in holographic form: the human biohologram.

Gariaev's research, if it is confirmed, opens the door to the next level of medical intervention. By accepting that our DNA has wave properties, it encourages us to look beyond the simple linear genetic model and to embrace other less invasive ways to combat disease and even prolong our lives. His idea that linguistic patterns are encoded fractally within the wave genome opens us to the possibility that specific sound, music, and talk therapies will evolve, with measurable outcomes achieved.

In addition, through an understanding of the role of light and interference patterns within our bodies, organs may be able to be regenerated without the need for hazardous surgery or complicated stem-cell technology. By studying the wave dynamics of viruses and bacteria, we will become open to developing new ways to manage viral illnesses, pandemics, and the bacterial diseases that persist particularly in Third World countries. As it is now widely recognized that we may be coming to the end of the antibiotic era,[106] it is essential that eyes are opened to this new paradigm.

Sadly, there are now reports that the work of Gariaev and his team of physicists has come to an abrupt halt at the hands of a highly skeptical Russian Academy of Sciences.[107] This group cites the research as being irrational and pseudoscientific. It is to be hoped that scientific skepticism is the real reason for this embargo; after all, the truth does have a way of eventually perfusing through even the most rigid of mind-sets. The possible alternative reason, that others with less benevolent intent are keen to suppress our knowledge of such matters, is an altogether more sinister scenario.

I would speculate, however, that in this subtle interactive domain of information fields and wave genetics, the noble intent of those conducting the research and of those performing the resulting clinical procedures will prove to be of paramount importance. Maybe we have not yet reached that point in our evolution at which this can be assured.

I hope, in my lifetime, to bear witness to this defining moment.

Chapter 23:
The Human Hologram—
Our Personal Renaissance

Be kind, for everyone you meet is fighting a hard battle.
—Plato (c. 428–348 BC), Greek philosopher

There have been times while writing this book that I have felt like a defense attorney, doggedly defending the right of my client, the human hologram, to be taken seriously. This would prove to be no easy task. For along with making a case to the judge and jury that there was more to my client than meets the eye, I would have to convince them that society would be all the better for knowing him. It would get harder still. Along the way, I would also have to sway them into considering that all they perceive, and have ever perceived, is but a virtual reality. That all the wonders of the beautiful spring morning they had encountered on their way to the courtroom—every bud on every tree, every bird, every courting couple walking hand in hand—all were but mere illusions, fanciful creative projections of their minds. Behind this facade, my case argument would continue, there really lies a fuzzy matrix, a mysterious unifying web connecting countless fields of information somehow existing outside time. I expect that, at this point of the proceedings, there would very likely be a concerted call for the defendant's eccentric hick-country lawyer to be locked away, together with his unwitting client, for life.

I have no doubt that the case for the prosecution would prove to be powerful. Whereas I would have reminded the court of the late Carl Sagan's words, "I believe that the extraordinary should be pursued," I strongly suspect that my learned colleague and adversary would take much glee in completing the quote: "But extraordinary claims require extraordinary evidence."

I am not sure how such a courtroom drama would be played out. Certainly, my case would be fiercely contested by a prosecution counsel representing influential clients who could well afford their healthy hourly rate of remuneration. However, I do believe that extraordinary evidence is accumulating to support a claim that we are indeed holographic beings. Our pursuit of this truth takes us down many roads, and possibly down several blind alleys too. But I suspect our diligence will be ultimately rewarded.

It appears that the most significant journey of discovery unfolds when we choose to venture within. The holographic paradigm supports this move, as by achieving a coherent state in our body, there is evidence we also induce coherence in others, both near and far. It is by going within that we exert changes in society as a whole. Thus social change becomes our own, not other people's responsibility.

And so this final chapter is dedicated to our own personal renaissance, as we strive to make practical use of the fresh insights we are gaining about ourselves and our universe. With this in mind, I have selected twelve important skills that each of us can learn to master, thereby ensuring a peaceful, stable, and joyful future for us all.

1. Trusting Our Feelings

Our feelings are our first and most reliable messages. When we are under threat, having to escape from danger, then feelings are immediately converted to actions as our survival is at stake. When we are called to rescue others from danger, then time is of the essence and should not be wasted analyzing our feelings or expressing our emotions.

As these episodes are, thankfully, rare for most of us, we all need training, with regular updating of our skills, so that we respond appropriately and effectively at times of crisis. We need to be fully prepared for natural disasters, and be adept in basic first-aid and resuscitation techniques. And to prevent our own stress levels rising after we have assisted others experiencing these traumatic events, it is important that we engage in effective debriefing, expressing our feelings honestly to others who listen intently.

At other times, however, it is prudent to ask ourselves if we are reacting appropriately. Are we overreacting to perceived threats that on reflection bear no threat to us at all? Are these feelings of being threatened so powerful and so frequent that other feelings of contentment, peace, and love are unable to break through?

If we are able to release unnecessary feelings of fear, so often conditioned into us in childhood, then we can begin to experience these more peaceful, but profound "fluctuations in the field." In a state of peace, we become open to inspiration and discover a source of creativity previously hidden to us.

2. Expressing Our Feelings

Expressing our feelings as quickly and honestly as possible is the next important step. The word "emotion" (e-motion) means "energy in motion." Letting others know our feelings is essential for maintaining our close relationships, and something that men, so conditioned to rescue, protect, fight, and escape, are generally still in the process of learning. As men, we are unaccustomed to expressing our feelings efficiently, and hence we tend to bottle up our anger, which eventually has to explode either outwardly or inwardly. Similarly, worries, if not shared, can compound and fester inside, making us depressed and sick.

As we become better at trusting and expressing our feelings, so we become better at listening to others. Listening to our spouse, partner, children, and friends is perhaps the most important preventative health-care intervention we can perform for them. Certainly, my own caseload would be halved, since my primary role as a doctor is as a listener to stories that should have been told a long time before. I strongly suspect that our emergency departments and hospitals would also notice a sustained lull if listening became a priority in the modern world. In addition, our skyrocketing pharmaceutical bill would begin to plummet.

3. Relating Rather Than Reacting

Expressing our feelings helps us relate to their meaning. It may well be that someone has annoyed us because their actions have reminded us of something we dislike about ourselves. In retrospect, we may recognize that this person is actually doing us a favor by, metaphorically speaking, holding a mirror up to our own behavior.

This process is facilitated by expressing any anger we may feel at the time, ideally aimed at the offensive act rather than at the offender. As a result, our anger will be resolved, and forgiveness for our adversary and ourselves can ensue. The human hologram model unites us all, and puts such conflicts into their true perspective.

4. Being Rational

As we grow to adulthood, our cognitive functions mature. We learn to process our feelings and emotions, we learn from our experiences, and we learn to appreciate the opinion of others. We release ourselves from dogma as we learn to honor those with different cultural backgrounds and religious beliefs. We discover our mental strengths, and our weaknesses, and become assertive while remaining humble. We learn to articulate our concerns, and express our frustrations, through an ever-expanding use of language. And we learn to solve our problems and the problems of others through actions performed for the greater good.

5. Acting with Integrity

We learn that through considerate acts of integrity, we free up our lives and the lives of others. There is no need for us to advertise our actions to the world; in fact, this would prove a waste of our valuable energy. The actions will broadcast themselves in a far more profound and sustaining way without any further effort. We can then reap the true rewards as, contented and at peace, we open ourselves automatically to yet more creative pursuits.

Similarly, all knowledge we gain in life should be passed on, ideally as soon as it is learned, as this benevolence will continue to keep us connected to a never-ending universal library of wisdom. The rewards of these actions, if we are patient, are so immense then it is no burden to take full responsibility for our lives in this way.

6. Showing Compassion to Ourselves

The responsible way is also the compassionate way. The art of self-compassion is perhaps the most difficult skill for us all to master. It is difficult to say no to someone without feeling guilty. Maybe we have been asked to a close friend's party, and yet every cell in our body yearns for an early night. Or maybe we have been asked by a friend to be involved in a campaign or cause that fails to ignite our heart. I continue to find these challenges the most testing of all. Saying no requires much skill. Perhaps the most helpful way is to not be forceful about one's own demands on others. When planning a dinner party, inquire first about your guest's busy life. Again, empathy will win out, friendships will grow, and embarrassing situations will diminish.

Holding compassion for ourselves—for instance, exhibiting a determination to heal—benefits others. Every week, I see grandmothers heal granddaughters, and vice versa, through becoming assertive in their acts of kindness to themselves. Ridding others of their long-held destructive feelings of guilt—their blame and their shame—is a valuable and lasting gift.

7. Showing Compassion to Others

Charity begins at home. It becomes easier to be kind to others if we are kind to ourselves. Then peace emanates from our being to be shared by others. We learn to love others as ourselves, because we all share so much. Modern genetics shows that all of us in this world have grown from common ancestors roaming the plains in East Africa; fifty thousand years ago our human population had shrunk to just a few thousand people. Considerably earlier on, approximately fourteen billion years ago, in the moments before the big bang, everything existed as an infinitely small dot; at this time before time, there was literally a zero degree of separation. And today, this state of connectivity persists in the form of entanglement. To quote the naturalist John Muir: "Tug on anything at all and you'll find it connected to everything else in the universe."

By creating peace within us, then there is evidence that our good wishes and prayers will transcend the barrier of space-time and be received wherever they are needed. Our acts of kindness, whether random or premeditated, whether great or small (there is really no difference), also penetrate the time barrier, fertilizing our present and our future.

But perhaps our sense of compassion for others is most easily accessed by realizing that every human on this earth faces significant challenges in this life. Or to paraphrase Plato's wise words that opened this final chapter: "We have all got it tough."

8. Coping with Chaos

We are here to be challenged. Life throws us problem after problem, demanding solution after solution. Life is unpredictable, thoroughly chaotic.

Underlying the structure of the universe may be a perfect symmetry, a heavenly choir singing in harmony, but there are times in all our lives when that beautiful noise is drowned out by something my schoolteachers used to call "an infernal din."

Those awkward, annoying moments that frustrate us so are as important as the blissful "in the zone" experiences that hold us in rapture. One such instance immediately springs to mind. On arriving at a secluded spot deep within the Whirinaki Forest Park in the center of the North Island of New Zealand, our qi-gong group assembled for a moment's quiet meditation to start our weeklong retreat. The very second we closed our eyes, a pneumatic drill burst into action less than thirty meters from our retreat site. The repair work on the road was essential for this impoverished region, which had been devastated some years earlier by the forced closure of a major timber mill. Although the roadwork continued for the duration of our stay, we learned to cope and to time our sessions around the noise.

Not every day will be happy. We won't pass every test, nor be good at everything. If everything went well, there would be no material for our comedians. So our daily mishaps not only teach us lessons, they also feed our sense of humor. The chaos inherent in our life on Earth reminds us that we are not always in control. It teaches us to embrace change and to welcome challenges.

If I was to create a recipe for a long and healthy life, then an ability to embrace change would be my prime ingredient. My ninety-four-year-old friend, New Zealand environmentalist John Hogan, is a living example of the benefits of this enduring human quality. During his amazing life, he contested a seat in parliament against a prime minister, became a long-time managing director of a science and technology museum, survived an early heart bypass operation thirty-five years ago, and over the past twenty years converted ten acres of suburban land back to pristine native bush. He recently he gifted this land, together with its six hundred magnificent kauri trees, to the local council, so that the community can benefit from all it offers for evermore.

9. Achieving Coherence

The ability to feel at peace is a special skill that, once learned, lasts a lifetime. In the human hologram model, this equates to the blissful feeling of being at one with the unified field. In this mode, we experience the lightness of being and the joy of the moment. It is a state of being I strive to achieve within myself during the second half of a consultation, after my initial attempts to sort out a rational plan of action.

I make it a personal priority to teach people seeking my help simple ways to achieve this state. (In appendix 1, I describe these exercises for your personal use.) In this state, our hearts beat more smoothly and efficiently, and our brains are more receptive to recalling memories and to creating ideas. Coherence within us individually spreads to others. And so, if we do these exercises in groups, they become simpler and more enjoyable, and their harmonious effects become more potent.

One famous example of coherence on a nonlocal scale is Lynne McTaggart's Intention Experiment where the growth of a plant has been shown to be enhanced by the focused coherent intent of thousands of people around the globe. In a similar vein, the Global Coherence Initiative, under the guidance of the HeartMath Institute, organizes regular synchronized events whereby participants from many countries achieve a coherent state together, while focusing their intent on healing and bringing peace to specific war-torn and disaster-stricken regions.

10. Examining Our Conditioned Beliefs

As we explore these personal commitments, we automatically question any unproductive beliefs held deeply within us. We may change our perception of the world, change our diet, or reinterpret spiritual or religious texts. For some, their faith will be deepened, as new meanings and insight emerge in the light of their expanded awareness. Others may feel a sense of alienation if friends and colleagues do not appear to be joining them on their voyage of discovery.

Some, to stay healthy and fulfilled, will challenge the message given to them in their childhood that they were "a born loser." Others will challenge the burden of responsibility loaded on them at an early age to "succeed at all costs" in an area of life that does not match their true vocation. Others still will realize they are trying hard to be perfect for the ultimate approval of others. Their health, and the health of future generations, depends on resolving these inner conflicts. We all have them, as will our children and our grandchildren.

It is clear that our evolution relies on the continuous questioning of our beliefs. For those who have suffered from the trauma of neglect, abuse, or abandonment, an awareness of how these events have influenced their health and their beliefs is important. Both their health and their beliefs can change for the better as a result, irrespective of their age or the extent of the trauma.

We are learning to challenge beliefs held dogmatically in our past. For example, there is now free discussion in most schools, and in most modern families, about sexual orientation. Homosexuality is no longer derided, and we are learning to respect and accept people for who they are, rather than sit in judgment over their sexual orientation. There is, of course, still much room for improvement here.

There has been an even greater reluctance in our society to discuss with freedom, and with any degree of scholarship, the subject of death and—a topic that, perhaps through fear, remains largely a subject of taboo. Death is often regarded as the failure of life, rather than its natural conclusion and its necessary partner. The commercial, material world focuses on youth, beauty, and physical strength rather than on the wisdom that evolves with age. Blockbuster movies, catering mainly to the young, often depict death graphically and violently, rather than serenely and peacefully. Lives are ended dramatically to the sound of gunfire, to the screech of brakes, or to an explosion that rocks the cinema-goers in their seats.

The human hologram, with its nonmaterial foundations, challenges this mind-set. During the writing of this book, my mother died peacefully in her ninetieth year. As families often do, we spent many hours poring over photographs of her life. When she was a baby, a toddler, a teenager, a physiotherapist in wartime India, a bride, a young mother, a working nurse/receptionist, a grandmother, and, for two months, a proud great-grandmother. Yet, despite nearly a century of ever-changing roles, and ever-changing body size and shape, she remained the same person with the same name, Pam. Pam was a permanent fixture inside an impermanent body. She remains a permanent fixture in the hearts of her loved ones.

In life, we perceive the world around us in a physical form that allows us to make a difference—something tangible for others who follow to build upon. In death, we leave behind this perception and the tools that make it happen. Every night, in sleep, we lay down these tools (our five senses) temporarily and suspend this perception; this represents a third of the time we are here on Earth, a full thirty years if we are fortunate enough to live a long life. It appears almost all living creatures need to take regular breaks away from their conscious activities if they are to lead sustainable lives. Bottlenose dolphins have to sleep half a brain (and half an eye) at a time, while the other half keeps them aware of predators and enables them to surface for breath. Fruit flies also sleep despite having the tiniest of brains, while research is under way on the sleep-like activities of jellyfish, with their very rudimentary nervous systems, and even the single-celled paramecium, which has no brain at all.[108] This research, together with the many studies listed in this book, is adding much to our knowledge about human and animal consciousness.

11. Being Advocates for Pure Science

Holographic science, fractal science, and quantum biology—we are only beginning to understand the immense significance of these new, largely inseparable, scientific disciplines. Their benefits for our species and our planet are likely to be vast—in fact, given our present limited state of knowledge, immeasurable. If we, as individuals, remain educated and watchful of the deeper implications of this science, it is highly likely that we will all benefit significantly in our lifetimes.

As our scientific knowledge expands to embrace realities beyond the purely physical and material, we need to remain cautious. Alongside the new science, there has to be a corresponding and parallel evolution of human values and ethics. So complex and technical are many of these advances that those involved in their intricacies may not have the broad wisdom or time to predict their true implications for humanity. Commercial interests dominate in science like never before, and more often than not, a modern scientific project is only viable because of the funding of a sponsor keen to develop a groundbreaking and lucrative product. With such a focus on material profit, there may be a tendency, either intended or otherwise, to downplay just how significant any findings are intrinsically for you and me, and for the generations that follow.

For example, if our brains and bodies are truly quantum processors, then the human being, aided but not dominated by machines, has a nearly unlimited potential to evolve consciously. Within our bodies and through our universal connections, we will be able to access endless resources to problem solve, and to heal. If our human brain as a quantum processor is a trillion times more powerful than originally thought, then imagine just what could be accomplished if all our planet's seven billion brains were synchronized as one within a state of perfect coherence.

And yet, if we glance at our daily news, our world appears far from coherent. It presents as being more chaotic, more complex than ever. In my own field of medicine, new illnesses born of this complexity are constantly emerging. Our immune systems and our mental processes are buckling under the weight of all this chaos and confusion. From 1996 to 2005, the number of Americans taking antidepressant medication doubled.[109]

Although antidepressants are widely reported to produce their effects by altering the levels of neurotransmitters between brain cells, there is new evidence from animal studies that these and other mood-altering medications also work within the cells, on the cytoskeleton.[110, 111] The cytoskeleton of our brains is now considered to play an essential role in the adaptation of the brain to change—a process now known as *neuroplasticity.* It is thought that the delay in clinical effects observed when starting a course of antidepressant drugs is due to this network gradually becoming influenced by the medication. As we have learned, the cytoskeleton of our brain cells is composed of various types of filament and of abundant microtubules. If microtubules prove to be the subcellular sites where our consciousness manifests out of quantum fields, then it will become clear that we are using medications that have their effects on our very perception of the world around us.

Whether this is an advantage or a disadvantage is up to the individual to decide, but one thing appears certain: we must be fully informed if our perceived reality is being altered through artificial means. And perhaps we should begin to consider as a species whether the underlying chaos that seems to be altering minds, in the form of depression and other mental illnesses, at such epidemic proportions is best managed by chemicals that potentially shift our consciousness in ways we do not yet understand. Or, instead, do we each need to address our lives, and our relationship with our world, at the most fundamental level, and make the changes we now know are possible?

And so it befits us all to become fully informed, and assertive, as science progresses to embrace matters of consciousness. With our vigilance, nanoscience and nanotechnology will be used safely for the greater good; I have met many in this field whose values give me real cause for hope. Ultimately, it is up to each and every one of us to ensure that this becomes so.

12. Being Responsible Parents and Teachers

If any of the messages I have tried to convey in this book have resonated with you—or perhaps have opened a line of inquiry or facilitated insights that have taken you beyond my simple explanations—then I encourage you to share this information with anyone who cares to listen. Allow free discussion, and healthy disagreement where appropriate—the more robust the debate the better. We continue to learn by changing and adapting our views by listening to others, thereby avoiding the trap of dogma. This is good science, and good sense. The shifts that are happening within us are also for the benefit of future generations. I have come to align the dynamics of healing with the dynamics of parenting, as this opens up the process to us all.

In this book, I have presented evidence that lends some credence to the theory that our universe, our world, and our body are bound together in a hidden form that defies our senses. I have explored the scientific evidence that supports this theory, together with a philosophical examination of how this knowledge, if it becomes accepted, might influence the human condition. The implications for our future, I believe, are profound.

We have yet to realize our full potential; the fractal, holographic quantum paradigm—the adjectives are almost interchangeable—opens us up to countless new possibilities. And yet it completely honors our physical presence here on Earth. In fact, we are in possession of the perfect tool—our body—to make the necessary changes that will ensure a sustainable future for us.

This century has seen a significant scientific breakthrough: we are now able to measure quantum processes in living tissue. We are able to do this for two reasons: (1) due to the quality of our education in science, math, and technology; and (2) due to the technology itself. We need, however, to instill one further vital ingredient if these advances are to represent a true paradigm shift. We need to apply a sense of values that can only be realized by examining our own selves. We need to understand how and why we feel the way we do. We need to recognize where we, as humans, have come from, and where we seek to go. The more of us who can find peace here on Earth, the more likely it is that a common vision will be created.

We are driven by the challenges presented to us in our complex, unpredictable lives. It is through meeting these challenges that we evolve. Each one of us is a unique variation on a theme; each of us a frantic, fractal whorl on the surface of a beautifully symmetrical Mandelbrot set. We work in the perfect medium to achieve our goals—a miracle of our perception. If we refer to the world around us as a virtual reality, it should in no way diminish its brilliance. A matrix of interference patterns, wondrous and mystical though it may be, does not presently excite, enthrall, or enrapture me like the warm glow of the setting sun, shared while basking in the arms of someone I love.

But appreciating a reality beyond our material existence allows us to gain true perspective on our lives. Peace and contentment can be valued alongside material comfort. Death may not hold such fear.

The signs of our holographic world are all around us, for each of us to savor. Beneath the chaos there lies a symmetry—life-affirming patterns that reverberate within our being and throughout the cosmos. We are here to be charmed by the heavenly and challenged by the hectic. The human hologram model is a work in progress. The philosophies and scientific papers presented here suggest that such a mind-stretching idea is far from preposterous; indeed, the time may have come to take it seriously.

Returning to my imaginary courtroom scene, I would now be drawing the case for the defense of my client, the human hologram, to a fitting close. I would naturally, first and foremost, thank the jury for following the argument so conscientiously. I would then remind them that my client's only alleged crime was one of loitering with benign intent in a state of harmony within the unified field. I would continue by expressing a hope that my recent line of reasoning would have encouraged them to look deeply inside themselves and thereby discover many noble traits that they share with the accused. And that by so recognizing the human hologram in their own being, as an act of kindness to one and all, the only verdict they could possibly reach would be a unanimous and resounding one leading to my client's eternal freedom.

At that point, as I do now, I would turn to the judge and say, "I rest my case."

Appendix I: Exercises

The universe consists entirely of waves of motion which spring from stillness and return to stillness.
—Walter Russell (1871–1963), American polymath

Walter Russell was describing the synchronous, symmetrical essence of our universe—the sustainable balance and order we associate with the first law of thermodynamics. We can apply Russell's vision to every facet of our personal lives, and to the cosmos as a whole. From a thought, a breath, a heartbeat, a human life, the swing of a pendulum, a symphony, a song, and a poem to the birth and death of a star or a galaxy. This is the true expression of the holographic universe, available for us all to experience every moment of our lives.

The exercises that follow aim to capture this essence—instilling a stillness and peace into our bodies that acts as a perfect antidote to the complexity, chaos, and overload that underlies so much of our ill health. We tend to fall sick because we neglect our basic needs whenever we are too busy, too worried, or too stressed.

The simplest, and to my mind the most profound, exercise involves an awareness of the act of breathing. I would advise every one of us to do this soon after we have climbed into bed at the end of the day. It need only take a few minutes.

1. Savoring the Breath

Lying on your back, in a perfectly relaxed state, allow your abdomen to rise with each in-breath, then fall with each out-breath. Allow your chest to follow your abdomen as your lungs fill while your diaphragm lowers.

The action is gentle and peaceful, never forced. Breathe slowly and mindfully in and out through both nostrils. The exercise is still effective if, like many consulting me, you have to breathe through your mouth because your nose is blocked!

The still-points between inspiration and expiration (and vice versa) are the most peaceful of all.

If your health is good, and your life is in balance, then this may be all that is needed.

You will find that you will be able to use this exercise throughout the day; a sense of calm can be achieved almost instantly.

2. Hands on Your Heart

Once a gentle breathing rhythm has been established, place your hands, one on top of the other, palm down on your sternum (breastplate). Now focus on the warmth in your chest created by your hands, as you continue the breathing exercise. Resist any temptation to rush. Allow yourself a subtle smile.

Your shoulders, arms, and hands now form a figure-eight shape. It may be helpful to visualize the Möbius heart diagram. (See figure 28.)

3. Words of Acceptance

You are now ready, if necessary, to use words, expressed either verbally or in your mind. You can use a familiar religious prayer if you wish. Focus not only on the meaning of the words, but also on their resonance—the vowel sounds in "love" and "God" are perfect matches. Allow the vowels to resonate gently within your chest.

This is the time to send your wishes to others, especially those who are unwell or having a difficult time. If you feel you have been dismissive or rude to anyone during the day, this is the time to wish them well.

Now is also the time to use any phrases specific to your healing. These are in a paired form, achieving a perfect balance between an acknowledgment of your feelings/symptoms and an expression of compassion and acceptance of your true self. For example you could say: "Even though I have this pain, I deeply and completely accept myself." It is ideal, however, if you are comfortable with this, to use the word "love" in the second part (see figure 41). It may be helpful to visualize the paired phrases within a figure-eight/infinity sign, with the meeting point resting in your heart.

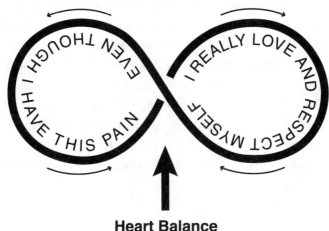

Heart Balance

Figure 41. Words of acceptance.

If it is appropriate, add a statement that may be relevant to your deeper healing. For example, "Even though I felt abandoned as a child, I really love and respect myself."

You may wish to add to this, possibly in consultation with a therapist. In my experience, it is best to keep the process very simple. Only move on to another phrase when you feel it is needed. For many of us, it takes time to feel comfortable in expressing compassion to ourselves. I recently received an e-mail from a young man with a diagnosis of bipolar disorder; he had noticed significant changes after six months of engaging in these exercises. In my view, the ritual is as important as brushing teeth; my dentist would not be too impressed if I taught my children to brush and floss for only two weeks out of every year!

4. Falling Asleep

You may now find you fall asleep with your hands over your heart. Sweet dreams!

It is my intention to teach others why this overall approach to our health is so important. To understand that we all need to examine our lives, and make the changes necessary so our health and joy is maximized. In addition, it helps to be listened to and to be understood. It helps to discuss the complexities of our lives and health with someone else. It helps to examine our conditioned beliefs together with someone who does not judge.

If we are to reverse the trend of ill health within our modern world, nothing in my experience can be more effective than a firm commitment to this inner inquiry. It is on this solid foundation that health professionals and doctors can begin to build their health plans with their clients and patients. This constitutes true ownership of one's health, and is as important for those requiring medication or surgery as it is for those undergoing counseling and natural therapies.

Appendix II: Experiments

IIA: Polarity and the Double Slit Quantum Eraser Experiment

We now have the technology to split photons of coherent laser light into two "entangled pairs," comprising complementary but opposing polarities. We can then examine whether, by somehow manipulating one beam, the other beam will respond accordingly. We also have equipment sensitive enough to record how these beams of light are behaving, that is, as bullets or as waves, and just how quickly changes in one beam may affect the other.

In 2002, physicists from Brazil published a landmark paper describing such an experiment, but with an added twist.(10) As in the experiment illustrated in figure 9, the scientists fired one of the paired beams of laser light through a quarter wave plate (QWP) on its way to two slits and a detecting screen. As they had expected, no interference pattern resulted. The next step was to examine whether altering the polarity of one beam could alter the behavior of the other.

To understand the rest of the experiment, a brief explanation about polarity is important. Polarization is the direction in which the electric field of the light is oscillating. There are two types of polarity: linear and circular. The linear type can either oscillate up and down (or the x direction) or to and fro (the y direction.) The circular type spins like a corkscrew, either counterclockwise to the left or clockwise to the right. (See figure 42.)

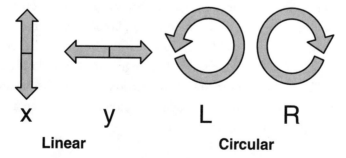

Figure 42. Polarization of light = the direction the field is oscillating.

In the experiment, a crystal split the light into two entangled beams of photons each with a linear-type polarity. It is known that when one photon in an entangled pair has an x type, the other always has a complementary y type (at 90 degrees to each other). This bond remains intact however far away they are from each other, effectively linking them together as a unit.

As previously described, one beam of light, in this case with y polarity, passed through the QWP, converting it to its circular polarity modes L and R, each of which then passed through their separate predictable slit, hence showing up as "bullets" on the screen. However, and this was the twist, the experimenters then disrupted the linear polarity of the other beam, placing a diagonal polarizer in its path that shifted its polarity to a 45-degree angle; that is, it was neither x nor y. This was instantly detected by the "entangled" beam going through the QWP and the slits, with the result that its own polarity changed reciprocally to a 45-degree form. As a result, rather than separating the beam neatly into clockwise R and counterclockwise L forms, which would pass through their separate slits, each half of the QWP produced a mixture of circular spins, making it impossible to predict which slit they were going through. Hence the bullet pattern disappeared (was erased!) and the interference pattern returned. (See figure 43.)

In the experiment, there was also a detector for the other beam (beam 1) lying behind the diagonal polarizer. For the final part of the experiment, they moved both the polarizer and detector further away, so beam 2 would encounter the QWP, slits, and its detector before beam 1 made contact with the polarizer (see figure 44). They repeated their measurements.

They discovered that the interference pattern had returned, showing in detector 2 before beam 1 had even reached the polarizer. And so somehow the photons in beam 2 seemed to predict what was coming. Just how did they know this? Entanglement, we know, connects photons outside the constraints of time and space. So could these photons be "conscious," even "super-conscious," possessed of the powers of precognition? Or could it be that an observer effect not only exists, but also functions outside the parameters of time and space? And is there a subtle interaction, a special entanglement, between the observers and the photons?

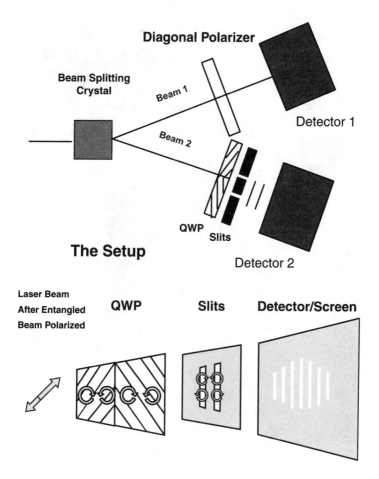

Figure 43. Changing (circular) polarization with quarter wave
plate (QWP). Every-which-way—observer can't predict.

Over the past decade, these questions have preoccupied many physicists.
In chapter 15, we explored theories that try to explain just how we perceive
light; and in appendix 2b, we examine experimental evidence that photons can
actually predict that a random outside influence (outside the experimenters'
control) will exert change on them before it actually occurs.

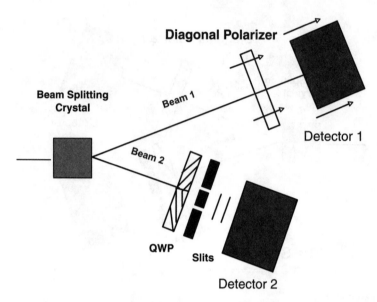

Figure 44. Moving the polarizer.

IIB: Experimental Realization of Wheeler's Delayed Choice Experiment 2007

In this experiment, photons were fired at a half-silvered mirror, a modern version of the "twin slits," splitting the beam in two.(105) A second beam splitter was placed 50 meters away; this could only be turned on and off completely randomly by a machine. (See figure 45.)

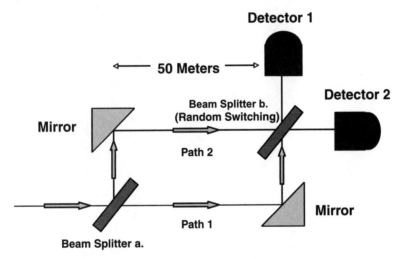

Figure 45. Wheeler's delayed choice experiment 2007.

When turned on, this second beam splitter recombined the split waves, and the resulting interference pattern showed up on the detectors in a typical "wave" form. When turned off, "bullet-like" photons were observed to hit each detector in equal numbers, showing that, as each individual bullet couldn't have been split, the light taking both path 1 and path 2 was in particle form from the start. Although defying logic, this means that the beam decides whether it is to behave as particles or waves for the full course of its journey, not at the start, but when the "system" decides later.

Endnotes

1. Nogier PMF, *Handbook to Auriculotherapy,* Moulins-les-Metz, France: Maisonneuve, 1981.

2. Niemtzow RC, Battlefield acupuncture, *Medical Acupuncture,* December 2007, 19(4):225–228; www.liebertonline.com/doi/abs/10.1089/ acu.2007.0603

3. Shonkoff JP, Boyce WT, McEwan BS, Neuroscience, molecular biology, and the childhood roots of health disparities, JAMA 2009;301:2252–2259.

4. Adverse Childhood Experiences Study, Centers for Disease Control and Prevention; www.cdc.gov/nccdphp/ace/about.htm

5. Usichenko TI, et al, Auricular acupuncture for pain relief after ambulatory knee surgery: a randomized trial. *Canadian Medical Association Journal* 2007;176:179–183; www.cmaj.ca/cgi/content/abstract/ 176/2/179

6. NIST physicists demonstrate quantum entanglement in mechanical system; www.nist.gov/public_affairs/releases/jost/jost_060309.html

7. www.symmetrymagazine.org/cms/?pid=1000198

8. Beckenstein JD, Information in the holographic universe, *Scientific American,* August 2003:59–65; www.scientificamerican.com/sciammag/ ?contents=2003-08

9. Chown M, Our world may be a giant hologram, *New Scientist,* January 15, 2009; 2691:24-27; www.newscientist.com/article/mg20126911. 300-our-world-may-be-a-giant-hologram.html

10. Walborn SP, et al, Double-slit quantum eraser, *Physical Review A,* February 20, 2002;65(3):033818.

11. Hyder SM, Schrödinger's cat experiment proposed, September 24, 2009; www.physorg.com/news173026471.html; based on: Towards quantum superposition of living organisms, http://arxiv.org/abs/0909.1469

12. Chown M, Could we create quantum creatures in the lab? *New Scientist,* September 15, 2009;19:39; www.newscientist.com/article/dn17792-could-we-create-quantum-creatures-in-the-lab.html

13. http://catholic-saints.suite101.com/article.cfm/saint-alphonsus-marie-liguori

14. Leiter D, The Vardøgr: perhaps another indicator of the nonlocality of consciousness, *Journal of Scientific Exploration* 2002;16(4):621–634.

15. Bem DJ, Honorton C, Does psi exist? Replicable evidence for an anomalous process of information transfer, *Psychological Bulletin* 1994;115:4–18.

16. Radin D, Analysis of ganzfeld experiments 1974–2004, *Entangled Minds,* New York: Paraview, 2006:120–121.

17. Schmidt S, et al, Distant intentionality and the feeling of being stared at: two meta-analyses, *British Journal of Psychology* 2004; 95:235–247.

18. Sherwood SJ, Roe CA, A review of dream ESP studies since the Maimonides dream ESP studies; in J Alcock, J Burns, and A Freeman (eds) *Psi Wars: Getting to Grips with the Paranormal Thorverton,* London: Imprint Academic, 2003.

19. Harris WS, et al, A randomized, controlled trial of the effects of remote, intercessory prayer on outcomes in patients admitted to the coronary care unit, *Arch Intern Med* 1999; 159:2273–2278.

20. Astin JA, et al, The efficacy of "distant healing": a systematic review of randomized trials, *Annals of Internal Medicine,* June 6, 2000; 132:903–910.

21. Benson H, et al, Study of the Therapeutic Effects of Intercessory Prayer (STEP) in cardiac bypass patients: a multicenter randomized trial of uncertainty and certainty of receiving intercessory prayer, *American Heart Journal,* April 2006;151(4):934–942.

22. Benedict C, Long-awaited medical study questions the power of prayer, *New York Times,* March 31, 2006; www.nytimes.com/2006/03/31/health/31pray.html

23. Roberts L, et al, Intercessory prayer for the alleviation of ill health, *Cochrane Database of Systematic Reviews,* April 15, 2009;15(2):CD000368.

24. Schnabel J, Remote Viewers: *The Secret History of America's Psychic Spies,* New York: Dell, 1997.

25. www.stat.ucdavis.edu/~utts/air2.html

26. http://en.wikipedia.org/wiki/Femtosecond

27. Engel GS, et al, Evidence for wavelike energy transfer through quantum coherence in photosynthetic systems, *Nature,* April 12, 2007;446:782–786.

28. Scholes GD, et al, Coherently wired light-harvesting in photosynthetic marine algae at ambient temperature, *Nature,* February 4, 2010;463:644–647.

29. www.smartplanet.com/business/blog/smart-takes/darpa-solicits-research-sensors-that-demonstrate-quantum-biology/5031. Also: www.fbo.gov/index?s=opportunity&mode =form&id=08a7fb6e82f0313a29227d05d0de6f71&tab=core&_cview=0

30. www.perfumerflavorist.com/fragrance/research/37572479.html

31. Turin L, A spectroscopic mechanism for primary olfactory reception, *Chemical Senses* 1996;21:773–791.

32. Brookes C, et al. Could humans detect odors by phonon assisted tunneling? *Physical Review Letters* 2007;98:038101.

33. Schulten K, et al, Magnetoreception through cryptochrome may involve superoxide, *Biophysical Journal,* June 17, 2009;96(12).

34. Cai J, et al, Quantum control and entanglement in a chemical compass, *Physical Review Letters* 2010;104:220502.

35. Tejero I, et al, Tunneling in green tea: understanding the antioxidant activity of catechol-containing compounds, Variational Transition-State Theory Study, *Journal of the American Chemical Society* 2007;129(18):5846–5854.

36. Fuxreiter M, Interfacial water as a "hydration fingerprint" in the noncognate complex of BamHI, *Biophysical Journal* 2005;89(2):903–911.

37. Pagnotta S, et al, Quantum behavior of water protons in protein hydration shell, *Biophysical Journal,* March 4, 2009; 96(5):1939–1943.

38. Schadschneider A, et al, Trafficlike collective movement of ants on trails: absence of jammed phase, *Physical Review Letters* 2009;102:108001.

39. Steck K, et al, Do desert ants smell the scenery in stereo? *Frontiers in Zoology* 2009;6(5); www.physorg.com/news187360582.html

40. Arias C, et al, Antibiotic-resistant bugs in the 21st century: a clinical super-challenge, *New England Journal of Medicine,* January 29, 2009;360(5):439–443.

41a. www.lifesci.ucsb.edu/~biolum/organism/milkysea.html

41b. www.nanotechwire.com/news.asp?nid=3020

42. Davies J, Everything depends on everything else, *Clinical Microbiology & Infection,* January 2009;15(1)Sup 1:1–4.

43. Paul Davies, a theoretical physicist, cosmologist, and astrobiologist who is leading the ASU cancer initiative; www.physorg.com/news175787974.html

44. Editorial review; www.amazon.com/Soul-White-Ant-Eugene-Marais/dp/0527612006

45. Lieberman-Aiden E, et al, Comprehensive mapping of long-range interactions reveals folding principles of the human genome, Science, October 9, 2009;326(5950): 289–293.

46. Langevin HM, Yandow JA, Relationship of acupuncture points and meridians to connective tissue planes, *Anatomical Record,* December 15, 2002;269(6):257–265.

47. Kim S, Coulombe PA, Emerging role for the cytoskeleton as an organizer and regulator of translation, *Nature Reviews Molecular Cell Biology,* January 2010;11:75–81.

48. Professor Popp is a member of the International Institute of Biophysics in Neuss, Germany. Comprehensive details on his and other biophysicists' research available at: www.lifescientists.de/index.htm

49. Ruth B, Popp FA, Experimental investigations on weak photoemission from biological systems (translated) *Z. Naturforschung* 1976;31c:741–745.

50. Popp FA, Ruth B, Bahr W, et al, Emission of visible and ultraviolet radiation by active biological systems, *Collective Phenomena* 3:187–214.

51. Schamhart DHJ, van Wijk R, *Photon Emission from Biological Systems,* ed. B Jezowska-Trzebiatowska, et al, Singapore: World Scientific, 1987:137–152.

52. Grasso F, et al, Photon emission from normal and tumor human tissues, *Experientia* 1992;48:10.

53. Kim J, et al, Measurements of spontaneous ultraweak photon emission and delayed luminescence from human cancer tissues, *Journal of Alternative and Complementary Medicine,* October 2005;11(5):879–884.

54. Popp FA, Gu Q, Li KH, Biophoton emission: Experimental background and theoretical approaches *Modern Physics Letters* 1994;B8:1269.

55. Reid B, On the nature of growth and new growth based on experiments designed to reveal a structure and function in laboratory space, parts 1 and 2, *Medical Hypotheses* 1989;29:105–144.

56. Wheeler J, Ford K, *Geons, Black Holes, and Quantum Foam: A Life in Physics,* New York: WW Norton, 1998.

57. Reid BL, Attempts to identify a control system for chemical reactions residing in virtual energy flows through the biosystem, *Medical Hypotheses* 1999;52:307–313.

58. Gariaev P, et al, Crisis in life sciences: the wave genetics response; http://www.emergentmind.org/gariaev06.htm

59. Bruza P, et al, Scientists model words as entangled quantum states in our minds, February 18, 2009, to appear in Proceedings of the Third Quantum Interaction Symposium, Lecture Notes in Artificial Intelligence, 2009, vol 5494.

60. Sun J, Deem MW, Spontaneous emergence of modularity in a model of evolving individuals, *Physical Review Letters* 2007;99:228107. See also www.physorg.com/news114185292.html

61. Gramling R, et al, Self-rated cardiovascular risk and 15-year cardiovascular mortality, *Annals of Family Medicine* 2008;6:302–306.

62. Disalvo D, Forget survival of the fittest: it is kindness that counts; www.scientificamerican.com/article.cfm?id=forget-survival-of-the-fittest.

    Keltner D, *Born to Be Good: The Science of a Meaningful Life,* New York: WW Norton, 2009.

63. Macey SL (ed), *Encyclopedia of Time,* New York: Garland, 1994:209.

64. www.nrao.edu/pr/2009/bhbulge

65. Irish L, et al, Long-term physical health consequences of childhood sexual abuse: a meta-analytic review, Journal of Pediatric Psychology 2010;35:450–461.

66. www.physorg.com/news126804909.html

67. Perkiömäki JS, et al, Fractal and complexity measures of heart rate variability, *Clinical and Experimental Hypertension* 2005;27(2–3):149–158.

68. Segerstrom SC, Solberg Nes L, Heart rate variability reflects self-regulatory strength, effort, and fatigue, *Psychological Science* 2007;18:275–281.

69. Sunkaria RK, et al, A comparative study on spectral parameters of HRV in yogic and non-yogic practitioners, *International Journal of Medical Engineering and Informatics* 2010;2(1):1–14; www.physorg.com/news176986454.html

70. McCraty R, Atkinson M, Tiller M, The role of physiological coherence in the detection and measurement of cardiac energy exchange between people, in Proceedings of the Tenth International Montreux Congress on Stress, Montreux, Switzerland, 1999.

71. www.heartmath.org/research/science-of-the-heart-head-heart-interactions.html

72. Stewart JC, et al, Depressive symptoms moderate the influence of hostility on serum interleukin-6 and C-reactive protein, *Psychosomatic Medicine* 2008;70:197–204.

73. Shen BJ, et al, Anxiety characteristics independently and prospectively predict myocardial infarction in men: the unique contribution of anxiety among psychologic factors, *Journal of the American College of Cardiology* 2008;51:113–119.

74. Davidson KW, et al, Don't worry, be happy: positive affect and reduced 10-year incident coronary heart disease: the Canadian Nova Scotia Health Survey, *European Heart Journal* 2010;31(9):1065–1070.

75. www.physorg.com/news9935.html

76. Nelson RD, Coherent consciousness and reduced randomness: correlations on September 11, 2001, *Journal of Scientific Exploration* 2002;16(4):549–570. See also: www.boundaryinstitute.org/bi/articles/AnomMag_web.pdf

77. Unique heart beat signature device could revolutionise healthcare, www.physorg.com/news183907385.html

78. Finoguenov A, et al, In-depth Chandra study of the AGN feedback in Virgo elliptical galaxy M84, *Astrophysical Journal* 2008;686(2):911.

    See also: www.sciencedaily.com¬ /releases/2008/11/081118161603.htm

79. Rosenthal JM, Okie S, White coat, mood indigo: depression in medical school, *New England Journal of Medicine*, September 15, 2005;353(11):1085–1088.

80. Manousakis E, Quantum formalism to describe binocular rivalry, *Biosystems,* November 2009;98(2):57–66.

81a. Reimers JR, et al, Study rules out Fröhlich condensates in quantum consciousness mode, 2009; www.physorg.com/news155904395.html

81b. Rahnama M, et al, Quantum collapse and visual consciousness, *NeuroQuantology* 2009;7(4):491–499.

82. Jones FW, Holmes DS, Alcoholism, alpha production and biofeedback, *Journal of Consulting and Clinical Psychology* 1976;44:224¬–228.

83. Pfurtscheller G, Event related desynchronization mapping: visualization of cortical activation patterns, in FH Duffy (ed), *Topographic Mapping of Brain Electrical Activity,* Boston: Butterworths, 1986:99–111.

84. Linsteadt S, Boekemeyer ME, *The Heart of Health: The Principles of Physical Health and Vitality,* Grass Valley, CA: Natural Healing House Press, 2003.

85. McCraty R, Atkinson M, Bradley RT, Electrophysiological evidence of intuition: part 1, the surprising role of the heart, *Journal of Alternative and Complementary Medicine* 2004;10(1):133–143. McCraty R, Atkinson M, Bradley RT, Electrophysiological evidence of intuition: part 2, a system-wide process, *Journal of Alternative and Complementary Medicine* 2004;10(2):325–336. Also McCraty R, et al, Coherent Heart 2009;5(2); www.integral-review.org/documents

86. Osborn J, Derbyshire S, Pain sensation evoked by observing injury in others, *Pain,* February 2010;148(2):268–274.

87. Jeon D, et al, Observational fear learning involves affective pain system and Cav1.2 Ca2+ channels in ACC, *Nature Neuroscience,* February 28, 2010;13:482–488.

88. Wells DL, Domestic dogs and human health: an overview, *British Journal of Health Psychology,* February 2007;12(1):145–156.

89. Keinan-Boker L, Vin-Raviv N, Liphshitz I, et al, Cancer incidence in Israeli Jewish survivors of World War II, *Journal of the National Cancer Institute,* 2009;101(21):1489–1500. Hursting SD, Forman MR, Cancer risk from extreme stressors: lessons from European Jewish survivors of World War II, *Journal of the National Cancer Institute,* 2009;101(21):1436–1437.

90. Wu M, et al, Interaction between Ras(V12) and scribbled clones induces tumour growth and invasion, *Nature,* January 28, 2010;463(7280):545–548.

91. Sood AK, et al, Adrenergic modulation of focal adhesion kinase protects human ovarian cancer cells from anoikis, *Journal of Clinical Investigation* 2010;120(5):1515–1523.

92. Kulik G, et al, Epinephrine protects cancer cells from apoptosis via activation of cAMP-dependent protein kinaseand BAD phosphorylation, *Journal of Biological Chemistry,* May 11, 2007;282(19):14094–14100.

93. Sprehn GC, et al, Decreased cancer survival in individuals separated at time of diagnosis: critical period for cancer pathophysiology? *Cancer,* November 1, 2009;115(21):5108–5116.

94. Church D, et al, Psychological symptom change in veterans after six sessions of emotional freedom techniques(EFT), 2009; www.wholistichealingresearch.com/91Church

95. www.usatoday.com/news/health/2010-02-04-health-care-costs_N.htm

96. Rowell HL, The impact of electronic media violence: scientific theory and research, *Journal of Adolescent Health* 2007;41(6 Suppl 1):S6–13.

97. Burundanga, *Wall Street Journal,* July 3, 1995; http://earthops.org/scopalamine1.html

98. Hughes H, et al, A study of patients presenting to an emergency department having had a "spiked drink," *Emergency Medicine Journal* 2007;24:89–91.

99. Moody R, Perry P, *Glimpses of Eternity,* New York: Guideposts, 2010.

100. Parnia S, et al, World's largest-ever study of near-death experiences; www.southampton.ac.uk/mediacentre/news/2008/sep/08_165.shtml

101. Spiegel D, et al, Effect of psychosocial treatment on survival of patients with metastatic breast cancer, Lancet 1989;2(86680):888–891.

102. Kurzweil R, *The Singularity Is Near: When Humans Transcend Biology,* New York: Penguin, 2006

103. www.wired.com/medtech/drugs/magazine/16-04/ff_kurzweil_sb

104. Liggins G, Howie R, A controlled trial of antepartum glucocorticoid treatment for prevention of the respiratory distress syndrome in premature infants, Pediatrics 1972;50:515–525.

105. Jacques V, Roch J-F, et al, Experimental realization of Wheeler's delayed-choice Gedanken experiment, Science, February 16, 2007;315(5814):966–968.

106. Alanis AJ, Resistance to antibiotics: are we in the post-antibiotic era? *Archives of Medical Research* 2005;36:697–705.

107. www.laleva.org/eng/2010/01/wave_genetics_research_targeted_by_russian_academy_skeptics.html

108. http://kibm.ucsd.edu/profile/2006-10/index.php

109. Olfson M, Marcus SC, National patterns in antidepressant medication treatment, *Archives of General Psychiatry* 2009;66(8):848–856.

110. Woolf N, et al, Impaired neuroplasticity and possible quantum processing derailment in microtubule, *NeuroQuantology,* March 2010;8(1):13–28.

111. Yang C, et al, Cytoskeletal alterations in rat hippocampus following chronic unpredictable mild stress and re-exposure to acute and chronic unpredictable mild stress, *Behavioural Brain Research* 2009;205(2):518–524.

Resources and Recommended Reading

Chown, Marcus. *Quantum Theory Cannot Hurt You: A Guide to the Universe.* London: Faber and Faber, 2007.

Church, Dawson. *The Genie in Your Genes: Epigenetic Medicine and the New Biology of Intention.* Santa Rosa, CA: Elite Books, 2009.

Dossey, Larry. *Healing Words: The Power of Prayer and the Practice of Medicine.* New York: HarperOne, 1995.

Gruder, David. *The New IQ: How Integrity Intelligence Serves You, Your Relationships, and Our World.* Santa Rosa, CA: Elite Books, 2008.

Hameroff, Stuart, Roger Penrose, David Chalmers, et al. *The Quantum Mind.* Memphis, TN: Books LLC, 2010; see also www.quantumconsciousness.org

Hubbard, Barbara Marx. *Conscious Evolution: Awakening Our Social Potential.* Novato, CA: New World Library,1998.

James, John. *The Great Field: Soul at Play in a Conscious Universe.* Santa Rosa, CA: Energy Psychology Press, 2007.

Kelly, Robin. *Healing Ways: A Doctor's Guide to Healing.* Auckland, NZ: Penguin, 2000.

———. *The Human Antenna: Reading the Language of the Universe in the Songs of Our Cells.* Santa Rosa, CA: Energy Psychology Press, 2010.

Lanza, Robert, and Bob Berman. *Biocentrism: How Life and Consciousness are the Keys to Understanding the True Nature of the Universe.* Dallas, TX: BenBella Books, 2009.

Lesmoir-Gordon, Nigel, and Will Rood. *Introducing Fractal Geometry.* London: Totem Books, 2001.

Lifton, Robert Jay. *The Nazi Doctors: Medical Killing and the Psychology of Genocide.* New York: Basic Books, 2000.

———. *Thought Reform and the Psychology of Totalism: A Study of "Brainwashing" in China.* Chapel Hill: University of North Carolina Press, 1989.

Lipton, Bruce. *The Biology of Belief: Unleashing the Power of Consciousness, Matter, and Miracles.* Carlsbad, CA: Hay House, 2008.

Lipton, Bruce, and Steve Bhaerman. *Spontaneous Evolution: Our Positive Future (and a Way to Get There from Here).* Carlsbad, CA: Hay House, 2009.

McTaggart, Lynne. *The Intention Experiment: Using Your Thoughts to Change Your Life and the World.* New York: Free Press, 2008.

Oschman, James. *Energy Medicine: The Scientific Basis.* New York: Churchill Livingstone, 2000.

Penrose, Roger. *Shadows of the Mind: A Search for the Missing Science of Consciousness.* New York: Oxford University Press, 1994.

Pribram, Karl. *Brain and Perception: Holonomy and Structure in Figural Processing.* Hillsdale, NJ: Lawrence Erlbaum Associates, 1991.

Radin, Dean. *The Conscious Universe: The Scientific Truth of Psychic Phenomena.* New York: HarperOne, 2009.

———. *Entangled Minds: Extrasensory Experiences in a Quantum Reality.* New York: Paraview, 2006.

Sheldrake, Rupert. *Morphic Resonance and the Presence of the Past.* Rochester, VT: Park Street Press, 1995.

Sidney Bender, Sheila, and Mary T. Sise. *The Energy of Belief: Psychology's Power Tools to Focus Intention and Release Blocking Beliefs.* Santa Rosa, CA: Energy Psychology Press, 2007.

Talbot, Michael. *The Holographic Universe.* New York: HarperCollins, 2006.

Velmans, Max. *Understanding Consciousness.* New York: Routledge, 2009.

Index

event horizon, 35, 36f7
evolution
    of consciousness, 91
    model, traditional, 88, 88f15b
    theory of, 66
exercises
    acceptance, words of, 222–23, 223f41
    breath, savoring the, 221–22
    falling asleep, 223–24
    hands on your heart, 222
explicate order, 27

F
Facebook, 26
FAK (enzyme), 165
Father Time, 92
fear
    about, 154, 179, 196–99, 210, 216, 220
    of death, 197–98, 220
    of dying, 141
    factor, 89–90
feelings. *See also* emotion
    of children, 143
    expressing our, 211
    heart-based, 29
    intuitive mind, 139f26
    move us into action, 29
    nonlocally connected heart to
        the unified field, 117
    puppy stoking, 153–54, 153f30
    rational mind, 139f27
    relating vs. reacting to, 21
    as subtle fluctuations in the field
        of consciousness, 146, 161
    trusting our, 210–11
femtosecond laser, 62, 198, 234
Fibonacci series, 28
fields. *See also* unified field
    biofields, 69–70, 98, 117
    of consciousness, 80, 83, 109
    harmonious, 14, 19, 19f2
    heart-generated electromagnetic,
        113, 117
    holographic, 79
    of information, 14, 71, 97–98, 129, 186
    of information, senses transform, 186
"fight or flight," 112
Finoguenov, Alexis, 118

First Law of Thermodynamics, 26, 195
Fitzgerald, F. Scott, 155
five element phases, 93f16
Fleming, Graham, 62–63
Flexitral, 64
forgiveness, 94, 108, 156, 211
Fourier analysis, 136
Fourier transform, 123, 124f20, 130
fractal(s)
    about, 14, 77, 83
    "blueprint," 14
    geometry, 73, 76, 98, 198
    human bodies, 75
    nature of image, 32
    nature of our universe, 53, 99
    pattern, self-replicating, 74–75
    pattern of life, 29
    patterns of time are, 93
    science, 217
fractal nature of human heartbeat, 117
free energy, 28
free will, 21, 80, 140, 179, 183, 185–86
friendship, 53
Fröhlich's condensates, 129
Fullerenes, 49n
functional MRI (fMRI) brain
        scans, 29, 135, 151
fundamentalist terror groups, 15
fungus gardens, 70

G
Gabor, Dennis, 13, 31, 34, 97, 122
Galileo, 13
gamma waves, 131, 131f23, 135
Gandhi, Mahatma, 149
ganzfeld telepathy, 55
Gariaev, Peter, 82, 98, 207
gene expression, 89f15c
gene replication, 90
generosity, 90
genes, 14
    epigenetic changes in, 106
    epigenetic factors acting on, 89, 89f15c
    for illness can lie dormant, 106
genetic code, 82
genetic engineering, 66
gene transfer, horizontal, 88f15b
genocide, 15